Interactions between Group Theory, Symmetry and Cryptology

Interactions between Group Theory, Symmetry and Cryptology

Special Issue Editor
María Isabel González Vasco

MDPI • Basel • Beijing • Wuhan • Barcelona • Belgrade

Special Issue Editor
María Isabel González Vasco
Universidad Rey Juan Carlos
Spain

Editorial Office
MDPI
St. Alban-Anlage 66
4052 Basel, Switzerland

This is a reprint of articles from the Special Issue published online in the open access journal *Symmetry* (ISSN 2073-8994) from 2018 to 2020 (available at: https://www.mdpi.com/journal/symmetry/special_issues/Group_Theory_Symmetry_Cryptology).

For citation purposes, cite each article independently as indicated on the article page online and as indicated below:

LastName, A.A.; LastName, B.B.; LastName, C.C. Article Title. *Journal Name* **Year**, *Article Number*, Page Range.

ISBN 978-3-03928-802-1 (Pbk)
ISBN 978-3-03928-803-8 (PDF)

Cover image courtesy of María Isabel González Vasco.

© 2020 by the authors. Articles in this book are Open Access and distributed under the Creative Commons Attribution (CC BY) license, which allows users to download, copy and build upon published articles, as long as the author and publisher are properly credited, which ensures maximum dissemination and a wider impact of our publications.

The book as a whole is distributed by MDPI under the terms and conditions of the Creative Commons license CC BY-NC-ND.

Contents

About the Special Issue Editor vii

Preface to "Interactions between Group Theory, Symmetry and Cryptology" ix

Jens-Matthias Bohli, María I. González Vasco, Rainer Steinwandt
Building Group Key Establishment on Group Theory: A Modular Approach
Reprinted from: *Symmetry* **2020**, *12*, 197, doi:10.3390/sym12020197 1

Yasir Nawaz and Lei Wang
Block Cipher in the Ideal Cipher Model: A Dedicated Permutation Modeled as a Black-Box Public Random Permutation
Reprinted from: *Symmetry* **2019**, *1485*, , doi:10.3390/sym11121485 12

Maria Bras-Amorós
Ideals of Numerical Semigroups and Error-Correcting Codes
Reprinted from: *Symmetry* **2019**, *11*, 1406, doi:10.3390/sym11111406 27

Kenneth Matheis, Rainer Steinwandt and Adriana Suárez Corona
Algebraic Properties of the Block Cipher DESL
Reprinted from: *Symmetry* **2019**, *11*, 1411, doi:10.3390/sym11111411 43

Sara D. Cardell, Verónica Requena, Amparo Fúster-Sabater and Amalia Orúe
Randomness Analysis for the Generalized Self-Shrinking Sequences
Reprinted from: *Symmetry* **2019**, *11*, 1460, doi:10.3390/sym11121460 59

Maria Bras-Amorós and Michael E. O'Sullivan
The Symmetric Key Equation for Reed–Solomon Codes and a New Perspective on the Berlekamp–Massey Algorithm
Reprinted from: *Symmetry* **2019**, *11*, 1357, doi:10.3390/sym11111357 85

María Cumplido, Juan González-Meneses and Marithania Silvero
The Root Extraction Problem for Generic Braids
Reprinted from: *Symmetry* **2019**, *11*, 1327, doi:10.3390/sym11111327 95

Jorge Martínez Carracedo
A Computational Approach to Verbal Width for Engel Words in Alternating Groups
Reprinted from: *Symmetry* **2019**, *11*, 877, doi:10.3390/sym11070877 110

Eligijus Sakalauskas, Aleksejus Mihalkovich
MPF Problem over Modified Medial Semigroup Is NP-Complete
Reprinted from: *Symmetry* **2018**, *10*, 571, doi:10.3390/sym10110571 122

José I. Escribano Pablos, María I. González Vasco, Ángel L. Pérez del Pozo, Misael E. Marriaga
The Cracking of *WalnutDSA*: A Survey
Reprinted from: *Symmetry* **2019**, *11*, 1072, doi:10.3390/sym11091072 135

About the Special Issue Editor

María Isabel González Vasco (Profesor Titular de Universidad) is an Associate Professor at MACIMTE, Universidad Rey Juan Carlos, where she has worked since 2003. She received her Diploma and Ph.D. degree in Mathematics from Universidad de Oviedo (1999 and 2003). Her research interests include provable security for cryptographic constructions, with a special focus on public-key cryptographic designs for encryption and group key exchange. She has published over 50 papers in the field, led two international research projects and acts regularly as a reviewer for several high-quality journals in the area as well as for top conferences. Further, she is involved in teaching related to mathematical cryptology at all levels. She is currently a member of the Board of Directors (Junta de Gobierno) of the Royal Spanish Mathematical Society.

Preface to "Interactions between Group Theory, Symmetry and Cryptology"

Cryptography lies at the heart of most technologies deployed today for secure communications. At the same time, mathematics lies at the heart of cryptography, as cryptographic constructions are based on algebraic scenarios ruled by group or number theoretical laws. Understanding the involved algebraic structures is, thus, essential to design robust cryptographic schemes.

This Special Issue is concerned with the interplay between group theory, symmetry and cryptography. It has been organized to highlight several exciting areas of research in which these fields intertwine: post-quantum cryptography, coding theory, computational group theory and symmetric cryptography. It is fair to say that all these areas are currently experiencing a resurgence, catalyzed by the urgent need for cryptographic solutions to resist quantum attacks.

Indeed, since the striking publication of Shor's quantum algorithms for factoring and computing discrete logarithms in polynomial time, the cryptographic community has searched for different, harder computational problems that can be used for cryptographic designs. In this book, three papers explore the computational hardness of certain group theoretical problems. In "The Root Extraction Problem for Generic Braids", by Cumplido et al., it is evidenced that finding the k-th root of an element in the braid group is generically fast, which, in particular, indicates the limitations of its cryptographic usage. On the other hand, the so-called MPF problem is proven NP-complete in the contribution of Sakalauskas et al., as a first step supporting its use for the construction and validation of related cryptographic primitives. Further, "A Computational Approach to Verbal Width for Engel Words in Alternating Groups", is concerned with a rewriting problem in alternating groups. Rewriting problems in non-abelian groups have inspired different cryptographic constructions since the eighties, and are still considered to be a promising source for hard computational problems.

Two papers contained in this issue are concerned with concrete cryptographic constructions for signature and key establishment. "The Cracking of WalnutDSA: A Survey", reviews the different attacks on a signature scheme, WalnutDSA, presented at the NIST standardization contest for post-quantum constructions. The security of WalnutDSA relies on certain rewriting problems over non-abelian groups, which have extensively been explored as a natural environment for quantum-resistant cryptographic primitives. The generic framework presented in the contribution "Building Group Key Establishment on Group Theory: A Modular Approach", by Bohli et al., aims at providing a sound design roadmap for the development of group key establishment protocols from group theoretical problems. Having quantum adversaries in mind, it seems worth exploring hard problems arising in non-abelian groups.

Coding theory is also understood as a potential arena for post-quantum cryptography. Two of the contributions in this Special Issue present recent relevant results in the field. The paper by Bras-Amorós and O'Sullivan establishes new results related to classical decoding algorithms used in public-key cryptography. Further, new fundamental relations between additive ideals of numerical semigroups and algebraic-geometry codes are presented in the contribution "Ideals of Numerical Semigroups and Error-Correcting Codes".

Finally, several works related to symmetric cryptography are contained in this volume. A nice algebraic analysis of DESL (a lightweight version of the block cipher DES) is given in "Algebraic Properties of the Block Cipher DESL" by Matheis et al., while the robustness of a pseudorandom number generator (used for the construction of stream ciphers) is explored in "Randomness Analysis for the Generalized Self-Shrinking Sequences" (Cardell et al.). In addition, the analysis of a block cipher modelled as a public random permutation is displayed in the contribution by Nawaz et al. These works evidence the usefulness of modelling and understanding the behaviour of symmetric tools and the permutations related to and induced by them.

María Isabel González Vasco
Special Issue Editor

Article

Building Group Key Establishment on Group Theory: A Modular Approach

Jens-Matthias Bohli [1], María I. González Vasco [2],* and Rainer Steinwandt [3]

1. Department of Information Technology, Mannheim University of Applied Sciences, 68163 Mannheim, Germany; j.bohli@hs-mannheim.de
2. MACIMTE, U. Rey Juan Carlos, 28933 Móstoles, Madrid, Spain
3. Department of Mathematical Sciences, Florida Atlantic University, Boca Raton, FL 33431, USA; rsteinwa@fau.edu
* Correspondence: mariaisabel.vasco@urjc.es

Received: 26 December 2019; Accepted: 19 January 2020; Published: 30 January 2020

Abstract: A group key establishment protocol is presented and proven secure in the common reference string mode. The protocol builds on a group-theoretic assumption, and a concrete example can be obtained with a decision Diffie–Hellman assumption. The protocol is derived from a two-party solution by means of a protocol compiler presented by Abdalla et al. at TCC 2007, evidencing the possibility of meaningfully integrating cryptographic and group-theoretic tools in cryptographic protocol design. This compiler uses a standard ring configuration, where all users behave symmetrically, exchanging keys with their left and right neighbor, which are later combined to yield a shared group key.

Keywords: group key establishment; group theory; provable security; protocol compiler

1. Introduction

Cryptography is the science of handling, storing, transmitting, and processing information securely, even in the presence of adversaries. For centuries, cryptographic techniques were developed for diplomatic or military scenarios, while nowadays individuals and institutions (often obliviously) make use of cryptographic tools every day. As a complex discipline, cryptography builds upon physics, mathematics, and different research areas within computer science. Mathematics are the main source of tools for cryptographic developments, in which, security is often demonstrated using the hardness of well understood mathematical problems. This paper is concerned with the construction of a widely used cryptographic tool, a group key exchange, using group theory as a base. Key exchange allows a number of users to establish a common secret value which will be subsequently used to secure their communication. Such cryptographic tools are often constructed from number theoretical problems (described in finite cyclic groups), and a challenging research question is whether secure constructions can be derived from different problems arising in group theory.

In recent years, not only due to the advent of quantum computation, significant efforts have been made to identify new mathematical platforms for implementing cryptographic schemes. One of the explored candidate platforms is the theory of finitely presented groups, where, in particular, a number of works on key establishment have been published. The first constructions in this direction where published about twenty years ago [1–3], and different approaches towards secure constructions have been explored regularly, such as [4–9], and the more recently [10]. Unfortunately, most of the proposed protocols have not been analyzed in a modern cryptographic security model (like [11–16]), and only few group-theoretic constructions with a rigorous security analysis seem to be known. This lack of formalism has resulted in weaknesses being overlooked (see, for instance, [17]).

One approach to facilitate the synergy between group-theoretic and cryptographic tools is the identification of general constructions that under suitable group-theoretic conditions yield an (efficient) cryptographic scheme with provable security guarantees. As examples for research along this line of thought, proposals for constructing IND-CCA secure asymmetric encryption schemes can be mentioned [18,19]. Also, constructions for building provably secure group key establishment schemes have been proposed (cf. [20,21]), but identifying practical non-abelian instances still appears to be a challenging problem. In this contribution, we build on [21], and try to extend and simplify their approach in the following sense:

- Instead of the random oracle model, we use the common reference string model. An (expected) price we pay for this, is the need of a decisional assumption instead of a computational one that is used in [21].
- Instead of setting out for a *group* key establishment directly, we suggest a construction for the two-party case and thereafter apply a protocol compiler of Abdalla et al. [22].

In terms of round complexity, we lose some efficiency through the modular design approach we chose. On the other hand, this modular design approach illustrates how an integration of group-theoretic and cryptographic tools can look like. Moreover, we obtain a comparatively clear group-theoretic condition which hopefully stimulates further research on finding concrete non-abelian instances. Concrete examples of our protocol can be derived from a decision Diffie–Hellman assumption, but we hope that in subsequent work also concrete non-abelian instances can be identified.

2. Preliminaries: Security Model and Protocol Goals

To explore the security of our protocol, we adopt the model used by Abdalla et al. [22], which can be traced back to [23–27]. Both to formulate our two-party solution and to use the "2-to-n compiler" from [22], we assume a common reference string (CRS) to be available that encodes the following information:

- Two values v_0, v_1. These will be the input for a pseudorandom function at the time of computing the session identifier and session key;
- The information necessary to implement a non-interactive and non-malleable commitment scheme (see Section 3.1 for further details);
- Two elements, chosen independently and uniformly at random, each taken from a family of universal hash functions (one as needed for the compiler in [22] and one for our two-party solution as detailed in Section 3.1).

This is similar to the constructions for password-authenticated key establishment in [24,27].

2.1. Communication Model and Adversarial Capabilities

As usual, we model protocol participants as probabilistic polynomial time (ppt) Turing machines (all our proofs hold for both uniform and non-uniform machines). We denote by $\mathcal{P}$ the total set of users which is assumed to be of polynomial size and by $\mathcal{U} = \{U_0, \ldots, U_{n-1}\} \subseteq \mathcal{P}$ the set of protocol participants. To enable authentication among the protocol participants, we assume that an existentially unforgeable signature scheme is available with all signing keys being chosen independently in a trusted initialization phase. The verification keys are assumed to be distributed in a trusted initialization phase, prior to the protocol execution.

2.1.1. Protocol Instances

We allow each protocol participant $U_i \in \mathcal{U}$ to execute polynomially many protocols *instances* in parallel. Each single instance $\Pi_i^{s_i}$ may be understood as a process executed by participant U_i. We will denote by $\Pi_i^{s_i}$ ($s_i \in \mathbb{N}$) the $s_i - th$ instance of user $U_i \in \mathcal{U}$, and the following seven variables are assigned to each instance:

used$_i^{s_i}$ will indicate whether this instance is or has been used for a protocol run. The used$_i^{s_i}$ flag is set through a protocol message received by the corresponding instance due to a call to the Send oracle (see below);

state$_i^{s_i}$ stores the state information needed during the protocol execution;

term$_i^{s_i}$ indicates if the execution has terminated;

sid$_i^{s_i}$ denotes a session identifier (which may be public) which may be later use as identifier for the session key sk$_i^{s_i}$ (in particular, the adversary is thus allowed to learn session identifiers);

pid$_i^{s_i}$ stores the user identities that $\Pi_i^{s_i}$ aims at establishing a key with. This set includes U_i himself;

acc$_i^{s_i}$ indicates that the protocol instance completed a protocol successfully. That is, whether the involved user accepted the session key or not;

sk$_i^{s_i}$ stores a distinguished NULL value in the beginning. After a session key is accepted by $\Pi_i^{s_i}$, this session key replaces the NULL value.

We refer to a paper of Bellare et al. [14] for more details on the usage of these variables.

2.1.2. Communication Network

The network is considered to be fully asynchronous and under complete control of the adversary. Arbitrary point-to-point connections among users are available, but the adversary may delay, eavesdrop, insert, and delete messages at will.

2.1.3. Adversarial Capabilities

We restrict to adversaries $\mathcal{A}$ running in probabilistic polynomial time, whose capabilities are made explicit through the four *oracles* listed below. These oracles formalize the interaction between $\mathcal{A}$ and the protocol instances run by the users. For the description of the Test oracle, we denote by b a bit that is chosen uniformly at random.

Send(U_i, s_i, M) This oracle sends a message M to instance $\Pi_i^{s_i}$ and returns the message generated by this instance. In case the instance $\Pi_i^{s_i}$ is previously unused and the message $M \subseteq \mathcal{P}$ contains a set of user identities, the used$_i^{s_i}$-flag is set, pid$_i^{s_i}$ initialized with pid$_i^{s_i} := \{U_i\} \cup M$. $\Pi_i^{s_i}$ initiates the protocol with the first message which is returned.

Reveal(U_i, s_i) This outputs the computed key of the instance stored in sk$_i^{s_i}$.

Test(U_i, s_i) If the corresponding session key is defined (i.e., acc$_i^{s_i}$ = true and sk$_i^{s_i} \neq$ NULL) and instance $\Pi_i^{s_i}$ is fresh (see Definition 4), $\mathcal{A}$ can execute this oracle query at any time when being activated. Then, if $b = 0$ the session key sk$_i^{s_i}$ is returned, while if $b = 1$ a uniformly chosen random session key is returned. An arbitrary number of Test queries is allowed for the adversary $\mathcal{A}$, but once the Test oracle returned a value for an instance $\Pi_i^{s_i}$, the same value will be returned for all instances partnered with $\Pi_i^{s_i}$ (see Definition 3).

Corrupt(U_i) This oracle models forward secrecy, as this query will output the secret signing key of user U_i.

2.2. Goals of a Key Establishment Protocol: Correctness, Integrity, and Security

We assume that an instance $\Pi_i^{s_i}$ always accepts the session key constructed at the end of a protocol run if no deviation from the protocol specification has occurred. The subsequent definition of correctness captures the protocol goal that, if the adversary is passive, all users involved in the same protocol session should come up with the same session key. By $\mathcal{A}$ being *passive*, we mean that $\mathcal{A}$ must not use the Corrupt oracle, and may query the Send oracle for the purpose of executing honest protocol executions only.

Definition 1 (Correctness). *A group key establishment protocol P is correct, if in the presence of a passive adversary $\mathcal{A}$ the following holds: for all i,j with both $\mathsf{sid}_i^{s_i} = \mathsf{sid}_j^{s_j}$ and $\mathsf{acc}_i^{s_i} = \mathsf{acc}_j^{s_j} = \mathrm{true}$, we have $\mathsf{sk}_i^{s_i} = \mathsf{sk}_j^{s_j} \ne \mathrm{NULL}$ and $\mathsf{pid}_i^{s_i} = \mathsf{pid}_j^{s_j}$.*

Unlike correctness, the concept of integrity imposes no restrictions on the adversary's behavior:

Definition 2 (Key Integrity). *A correct group key establishment protocol fulfills key integrity, if all instances of users that have accepted with the same session identifier $\mathsf{sid}_j^{s_j}$ hold with overwhelming probability identical session keys $\mathsf{sk}_j^{s_j}$ and identical partner identifiers $\mathsf{pid}_j^{s_j}$.*

Finally, for defining security, we detail our interpretation of partnering and freshness:

Definition 3 (Partnering). *Instances $\Pi_i^{s_i}$ and $\Pi_j^{s_j}$ are partnered if $\mathsf{pid}_i^{s_i} = \mathsf{pid}_j^{s_j}$, $\mathsf{sid}_i^{s_i} = \mathsf{sid}_j^{s_j}$, and $\mathsf{acc}_i^{s_i} = \mathsf{acc}_j^{s_j} = \mathrm{true}$.*

The idea of freshness is to characterize those instances where the adversary does not know the secret session key for trivial reasons. In particular, note that after revealing a session key from instance $\Pi_i^{s_i}$, the session keys of all instances partnered with $\Pi_i^{s_i}$ are known, too:

Definition 4 (Freshness). *An instance $\Pi_i^{s_i}$ is called* fresh *provided that none of the following condition holds:*

- *For some $U_j \in \mathsf{pid}_i^{s_i}$ a query $\mathsf{Corrupt}(U_j)$ was executed before a query of the form $\mathsf{Send}(U_k, s_k, *)$ has taken place where $U_k \in \mathsf{pid}_i^{s_i}$.*
- *The adversary queried $\mathsf{Reveal}(U_j, s_j)$ with $\Pi_i^{s_i}$ and $\Pi_j^{s_j}$ being partnered.*

Now the advantage $\mathsf{Adv}_{\mathcal{A}}(\ell)$ of a probabilistic polynomial time adversary $\mathcal{A}$ in attacking a key establishment protocol P is the function

$$\mathsf{Adv}_{\mathcal{A}} := |2 \cdot \mathsf{Succ}_{\mathcal{A}} - 1|$$

in the security parameter ℓ. Here, $\mathsf{Succ}_{\mathcal{A}}$ denotes the probability that $\mathcal{A}$ queries Test only on fresh instances and correctly outputs the bit b used by the Test oracle while preserving the freshness of all instances queried to Test.

Definition 5. *We say that an authenticated group key establishment protocol P is* secure, *if the following inequality holds for every probabilistic polynomial time adversary $\mathcal{A}$ some negligible function $\mathsf{negl}(\ell)$ in the security parameter ℓ: $\mathsf{Adv}_{\mathcal{A}}(\ell) \le \mathsf{negl}(\ell)$*

As in [22], our security definition above implies forward secrecy. Specifically, our freshness definition (Definition 4) allows Test queries to an instances, for which the long term secret key has been revealed by a Corrupt query (or is partnered with a instance that has be queried Corrupt) as long as the adversary has not asked a Send query to any of these instances (or their partners) after the Corrupt query.

3. Building on a Group-Theoretic Assumption

As already indicated, we construct our group key establishment protocol in two steps: In Section 3.1 we describe a two-party solution, which subsequently is lifted to an n-party solution by means of the protocol compiler in [22].

3.1. A Two-Party Solution

On the cryptographic side, our two-party solution mainly builds on three technical tools:

- A **non-interactive non-malleable commitment scheme** $\mathcal{C}$, satisfying the following requirements:
  - It is *perfectly binding* in the sense that every commitment can be decommitted to at most one value.
  - It is *non-malleable for multiple commitments*. This means that an adversary who knows commitments to a polynomial sized set of values ν, will not be able to output commitments to a polynomial sized set of values β related to ν in a meaningful way. It is well-known that in the CRS model such a commitment scheme can be implemented by means of any IND-CCA2 secure public key encryption scheme, for instance.
- A **family of universal hash functions** $\mathcal{UH}$ mapping triples consisting of two elements from G and a $\text{pid}_i^{s_i}$-value onto a superpolynomial sized set $\{0,1\}^L$. A universal hash function UH will be selected by the CRS from this family.
- A **collision-resistant pseudorandom function family** $\mathcal{F} = \{F^\ell\}_{\ell \in \mathbb{N}}$ (see Katz and Shin [28]). We assume $F^\ell = \{F_\eta^\ell\}_{\eta \in \{0,1\}^L}$ to be indexed by $\{0,1\}^L$ and further denote by $v_0 = v_0(\ell)$ a publicly known value such that no ppt adversary can find two different indices $\lambda \neq \lambda' \in \{0,1\}^L$ such that $F_\lambda(v_0) = F_{\lambda'}(v_0)$. We further use another public value v_1, fulfilling the same requirement as v_0 for deriving the session key (this can also be included in the CRS—see [28] for more details).

Our protocol builds on [21], and for the security proof we have to assume that the underlying group G (respectively, the family of groups $G = G(\ell)$, indexed by the security parameter) satisfies a number of conditions. Besides assuming products and inverses of group elements to be computable by efficient (ppt) algorithms, we further assume G to have a ppt computable canonical representation of elements. The latter allows us to identify group elements with their canonical representation. Furthermore, as in [21], we need three algorithms to perform the computations occurring in a protocol execution:

- DomPar, the *domain parameter generation* algorithm, is a (stateless) ppt algorithm that, upon input of the security parameter 1^ℓ, outputs a finite sequence S of elements in G. The subgroup of G spanned by S, $\langle S \rangle$, will be publicly known. Note that, for the special case of applying our framework to a DDH-assumption, S specifies a public generator of a cyclic group.
- SamAut, the *automorphism group sampling* algorithm, is a (stateless) ppt algorithm that, upon input of the security parameter 1^ℓ and a sequence S output by DomPar, returns a description of an automorphism ϕ on the subgroup $\langle S \rangle$, so that both ϕ and ϕ^{-1} can be efficiently evaluated. For example, for a cyclic group, ϕ could be given as an exponent, or for an inner automorphism the conjugating group element could be specified.
- SamSub, the *subgroup sampling* algorithm, is a (stateless) ppt that, upon input of the security parameter 1^ℓ and a sequence S output by DomPar, returns a word $x(S)$ representing an element $x \in \langle S \rangle$. Intuitively, SamSub chooses a random $x \in \langle S \rangle$, so that it is hard to recognize x if we know elements of x's orbit under $\text{Aut}(\langle S \rangle)$. Thus, our protocol requires an explicit representation of x in terms of the generators S.

With this notation, we can now define a decision problem, whose supposed difficulty will be essential for our security proof. As usual, with the notation $o \leftarrow A(i)$ we describe that algorithm A upon receiving input i outputs o:

Definition 6 (Decision Automorphism Application). *Suppose that we have fixed a quadruple $(G, \mathsf{DomPar}, \mathsf{SamAut}, \mathsf{SamSub})$. Then the decision automorphism application (DAA) assumption states that for all ppt algorithms $\mathcal{A}$ the advantage function $\mathsf{Adv}_{\mathcal{A}}^{\mathsf{DAA}} = \mathsf{Adv}_{\mathcal{A}}^{\mathsf{DAA}}(\ell) :=$*

$$\left| \Pr\left(\mathcal{A}(S, x, (\phi_i(S), \phi_i(x))_{i=1,2}) = 0 \;\middle|\; \begin{array}{l} S \leftarrow \mathsf{DomPar}(1^\ell), \quad x \leftarrow \mathsf{SamSub}(1^\ell, S), \\ (\phi_i \leftarrow \mathsf{SamAut}(1^\ell, S))_{i=1,2} \end{array} \right) - \right.$$
$$\left. \Pr\left(\mathcal{A}(S, r, (\phi_i(S), \phi_i(x))_{i=1,2}) = 0 \;\middle|\; \begin{array}{l} S \leftarrow \mathsf{DomPar}(1^\ell), \quad x \leftarrow \mathsf{SamSub}(1^\ell, S), \\ (\phi_i \leftarrow \mathsf{SamAut}(1^\ell, S))_{i=1,2}, \quad r \leftarrow \mathsf{SamSub}(1^\ell, S) \end{array} \right) \right|$$

is negligible.

Example 1 (Building on decision Diffie–Hellman). *Let G be a finite cyclic group and $S := \langle g \rangle$ a prime order subgroup with generator g of order q. If we let SubSam choose uniformly at random an exponent $x \in \{1, \ldots, q-1\}$ and SamAut uniformly at a random exponent $\phi \in \{1, \ldots, q-1\}$, then the DAA problem just described can be recognized as polynomial-time equivalent to a decision Diffie–Hellman (DDH) problem:*

"**DDH solution $\Rightarrow$ DAA solution**": *When facing, the DAA problem, we obtain as input a tuple $(g, g^y, (g^{\phi_i}, g^{x\phi_i})_{i=1,2})$ where either $y = x$, or y has been chosen uniformly at random from $\{1, \ldots, q-1\}$—independently of x and the ϕ_is. Given a DDH oracle, we just query it with $(g, g^y, g^{\phi_1}, g^{x\phi_1})$ to see with non-negligible success probability which is the case.*

"**DDH solution $\Leftarrow$ DAA solution**": *When facing the DDH problem, we obtain as input a tuple $(g, g^{\phi_1}, g^x, g^y)$, where either $y = \phi_1 x \bmod q$, or y has been chosen uniformly at random from $\{1, \ldots, q-1\}$—independently of x and ϕ_1. Choosing another random $\phi_2 \in \{1, \ldots, q-1\}$, we can compute the input*

$$(g^{\phi_1}, g^y, ((\underbrace{g}_{=(g^{\phi_1})^{\phi_1^{-1}}}, \underbrace{g^x}_{=(g^{\phi_1 x})^{\phi_1^{-1}}}), (\underbrace{g^{\phi_2}}_{=(g^{\phi_1})^{\phi_1^{-1}\phi_2}}, \underbrace{(g^x)^{\phi_2}}_{=(g^{\phi_1 x})^{\phi_1^{-1}\phi_2}})))$$

needed for a DAA attacker. Running a successful DAA attacker with this input, we immediately obtain the desired DDH attacker.

A two-party key establishment protocol building on the DAA assumption is presented in Figure 1. The figure describes the operations to be performed by instance $\Pi_i^{s_i}$ of U_i. For the sake of readability we name the users trying to establish a common key as U_0 and U_1, and here, as in the sequel, we often omit making explicit the identifiers s_i of the instances $\Pi_i^{s_i}$ involved in the protocol execution and just write sid_i instead of $\mathsf{sid}_i^{s_i}$, for instance. The common reference string is denoted by ρ, and for a commitment to a value x involving random choices r we write $C_\rho(x; r)$. Finally, S denotes the subgroup generators which are to be fixed prior to the protocol execution by means of DomPar (and may also be included in the CRS ρ).

In the subsequent section we prove the following result:

Proposition 1 (Security of the Two-Party Protocol). *Assume that for each ppt time algorithm $\mathcal{A}$, its advantage $\mathsf{Adv}_{\mathcal{A}}^{\mathsf{Sig}}$ of achieving an existential forgery under the adaptive chosen-message attack for the underlying signature scheme, and $\mathsf{Adv}_{\mathcal{A}}^{\mathsf{DAA}}$, its advantage of solving DAA, can be bounded by a negligible function (in ℓ). Then the protocol in Figure 1 is a correct and secure two-party key establishment protocol fulfilling key integrity.*

In Figure 2, we describe the group key establishment protocol obtained from a given two party group key establishment protocol 2-AKE via the compiler from [22]. We note here that given the result of Proposition 1, we can apply [22, Theorem 1] (which, as noted by Nam et al. in [29] is only valid if the underlying two party construction fulfills integrity) to obtain our desired security result:

Corollary 1 (Security of the *n*-Party Protocol). *Denoting the two-party key establishment protocol in Figure 1 by 2-AKE, the protocol described in Figure 2 is a secure group key establishment fulfilling key integrity.*

Round 1:

Initialization: For $i = 0, 1$ the variables of involved oracles $\Pi_i^{s_i}$ are set as $\text{pid}_i := \{U_0, U_1\}$, $\text{used}_i := \text{true}$.
Also, for $i = 0, 1$, choose $(\phi_i, (\phi_i)^{-1}) \leftarrow \mathsf{SamAut}(1^\ell, S)$, $x_i \leftarrow \mathsf{SamSub}(1^\ell, S)$.
Computation: User U_i, for $i = 0, 1$ chooses a random r_i and constructs a commitment
$$c_i := C_\rho(x_i; r_i).$$

Communication: User U_i, $i = 0, 1$, sends $m_i^1 := (U_i, \phi_i(S), c_i)$ to U_{1-i}.

Round 2:

Computation: User U_i, $i = 0, 1$, computes $\phi_{1-i}(x_i)$ and a signature σ_i of $(U_i, \phi_{1-i}(x_i))$ (using the representation of $x_i = x_i(S)$ in terms of the generators S and the images $\phi_i(S)$ of the subgroup generators).

Communication: Each user U_i, $i = 0, 1$, sends $m_i^2 := (U_i, \phi_{1-i}(x_i), \sigma_i)$ to U_{1-i}.

Key Generation:

Computation: Compute x_{1-i} by applying $(\phi_i)^{-1}$ to $\phi_i(x_{1-i})$, and define the master key
$$K := (x_0, x_1, \text{pid}_i).$$

Verification: Check the correctness of the commitment c_{1-i} and the signature σ_{1-i}.
If true, set $\text{sk}_i := F_{\text{UH}(K)}(v_1)$, $\text{sid}_i := F_{\text{UH}(K)}(v_0)$ and $\text{acc}_i := \text{term}_i := \text{true}$.
Else set $\text{acc}_i := \text{false}$, $\text{term}_i := \text{true}$

Figure 1. A two-party key establishment protocol in the common reference string (CRS) model.

Round 0:

2-AKE: For $i = 0, \ldots, n-1$ execute 2-AKE(U_i, U_{i+1}), (where, as customary, all indices are to be taken mod n, i.e., $U_n = U_0$, etc.).
Thus, each user U_i holds two keys $\overrightarrow{K}_i$, $\overleftarrow{K}_i$ shared with U_{i+1} respectively U_{i-1} and (non-secret) corresponding session identifiers $\overrightarrow{\text{sid}}_i$, $\overleftarrow{\text{sid}}_i$.

Round 1:

Computation: Each U_i computes
$$X_i := \overrightarrow{K}_i \oplus \overleftarrow{K}_i$$
and chooses a random r_i to compute a commitment $C_i = C_\rho(U_i, X_i; r_i)$.

Broadcast: Each U_i broadcasts $M_i^1 := (U_i, C_i)$.

Round 2:

Broadcast: Each U_i broadcasts $M_i^2 := (U_i, X_i, r_i)$.

Check: Each U_i checks that $X_0 \oplus X_1 \oplus \cdots \oplus X_{n-1} = 0$ and the correctness of the commitments.

Computation: Each U_i sets $K_i := \overleftarrow{K}_i$ and computes the $n-1$ values
$$K_{i-j} := \overleftarrow{K}_i \oplus X_{i-1} \oplus \cdots \oplus X_{i-j} \quad (j = 1, \ldots, n-1),$$
defines a master key
$$K := (K_0, \ldots, K_{n-1}, \text{pid}_i),$$
and sets $\text{sk}_i := F_{\text{UH}(K)}(v_1)$, $\text{sid}_i := F_{\text{UH}(K)}(v_0)$ and $\text{acc}_i := \text{true}$.

Figure 2. The protocol compiler from [22].

3.2. Security Analysis for the Two-Party Case: Proof of Proposition 1

Correctness and Integrity. Due to the collision-resistance of the family $\mathcal{F}$, all oracles that accept with identical session identifier use the same index value $UH(K)$ and therewith also obtain the same session key and have identical pid_i-values with overwhelming probability.

Security. Let q_s and q_t denote the (polynomially bounded) number of adversarial queries to the Send and Test oracle, respectively.

We consider a simulator simulating all oracles and instances for the adversary. The proof is thus set up following a sequence of experiments or games, where from game to game the simulator's behavior deviates from the previous in a certain controlled way. We follow standard notation and we denote by $\mathsf{Adv}(\mathcal{A}, G_i)$ the advantage of the adversary when confronted with Game i and by $\mathsf{Succ}(\mathcal{A}, G_i)$ the success probability of $\mathcal{A}$ winning in Game i. As usual, the security parameter will be denoted denoted by ℓ.

Game 0. All oracles are simulated as defined in the model. Thus, $\mathsf{Adv}(\mathcal{A}, G_0)$ is exactly $\mathsf{Adv}_{\mathcal{A}}$ and $\mathsf{Succ}(\mathcal{A}, G_0)$ is the probability of violating the security of our key exchange protocol.

Game 1. In this game, the simulator keeps a list with entries (i, M, σ_M) for every message M and corresponding signature σ_M he has produced and returned to the adversary $\mathcal{A}$ in a Round 2 message following a Send query.

By Forge we denote the event that $\mathcal{A}$ queries the Send oracle with a message M containing a valid signature σ_M of an uncorrupted principal U_i and with (i, M, σ_M) not being contained in the simulator's list. If the event Forge occurs, we abort the simulation and take the adversary $\mathcal{A}$ for being successful in breaking the security of the protocol. Thus,

$$|\mathsf{Succ}(\mathcal{A}, G_1) - \mathsf{Succ}(\mathcal{A}, G_0)| \leq P(\mathsf{Forge}) \qquad (1)$$

Lemma 1. *If the signature scheme used in the above protocol is existentially unforgeable under adaptive chosen-message attacks, then $P(\mathsf{Forge})$ is negligible:* $P(\mathsf{Forge}) \leq |\mathcal{P}| \cdot \mathsf{Adv}_{\mathcal{A}}^{\mathsf{Sig}}$.

Proof. Any ppt adversary $\mathcal{A}$ provoking the event Forge can be turned into an attacker against the underlying signature scheme by means of our simulator: The simulator obtains the public verification key PK and access to a signing oracle. In the initialization phase of the protocol, the simulator assigns the key PK uniformly at random to one of the at most $|\mathcal{P}|$ users the adversary can involve. Whenever during the subsequent simulation a signature for this user has to be generated, the simulator queries the signing oracle.

If $\mathcal{A}$ comes up with a message/signature pair that is not stored in the simulator's list, the simulator returns this message as existential forgery. If $\mathcal{A}$ does not come up with such a message, the simulator outputs $\bot$. Having chosen the party U_i uniformly at random, the simulator's success probability for an existential forgery is at least $1/|\mathcal{P}| \cdot P(\mathsf{Forge})$, and we get $P(\mathsf{Forge}) \leq |\mathcal{P}| \cdot \mathsf{Adv}_{\mathcal{A}}^{\mathsf{Sig}}$. □

Thus, from Equation (1), we get

$$|\mathsf{Adv}(\mathcal{A}, G_1) - \mathsf{Adv}(\mathcal{A}, G_0)| \leq \mathsf{negl}(\ell) \qquad (2)$$

Game 2. Now the simulation of the Test oracle is modified, so that, on input of a fresh instance, it will always output an element selected uniformly at random in the key space. Thus, $\mathsf{Adv}(\mathcal{A}, G_2) = 0$.

Suppose that $\mathcal{A}$ is able to distinguish between Game 2 and Game 1. We construct an attacker D, that breaks the DAA assumption and uses $\mathcal{A}$ as a black-box. The attacker D will start by setting up the instances with key pairs for the signature scheme and receive a DAA-instance as a challenge. Further, D will choose an index $a \in \{1, \ldots, q_t\}$ uniformly at random and select two values $u, v \in \{1, \ldots, q_s\}$ chosen independently and uniformly at random subject to the condition $u \neq v$. Then the adversary

$\mathcal{A}$ is started. D will simulate the model as in Game 1 except for the uth and vth instance activated by the adversary $\mathcal{A}$ and the answers to the Test query. For the uth and vth instances activated by $\mathcal{A}$, the messages will be constructed from the DAA challenge. If these two instances do not end up in the same session, D aborts the simulation and starts anew. The same happens, if $\mathcal{A}$ does not query his ath Test query to one of these two instances.

D will simulate the Test oracle as follows: The first $a-1$ queries of Test will be answered with the real session key, in the ath query, D will return the challenge, and from query $a+1$ on, D will always answer with a random element.

By a standard hybrid argument, D will win the challenge in $1/q_t$ of the cases where $\mathcal{A}$ distinguished Game 1 and Game 2. Excluding the necessary aborts (namely, if the instances that were chosen were not those used in the ath query of Test), we have:

$$|\text{Adv}(\mathcal{A}, G_2) - \text{Adv}(\mathcal{A}, G_1)| \leq q_s^2 q_t \text{Adv}^{\text{DAA}} \tag{3}$$

Combining Equations (2) and (3) yields the desired negligible upper bound for $\text{Adv}_\mathcal{A}$.

4. Conclusions

Our discussion evidences the possibility of meaningfully integrating tools from group theory and cryptography. Unfortunately, so far we cannot provide a concrete non-abelian example, but a concrete instance of our protocol can be derived by means of the decision Diffie–Hellman assumption. We hope, however, that the modular approach taken above facilitates the design of group key establishment schemes building on group-theoretic tools and fertilizes the exchange of ideas between group theory and cryptography.

Author Contributions: All authors contributed equally to this paper, and were cooperatively involved in conceptualization, investigation, formal analysis and writing. All authors have read and agreed to the published version of the manuscript.

Funding: This research was sponsored in part by the NATO Science for Peace and Security Programme under grant G5448 and in part by Spanish MINECO under grant MTM2016-77213-R.

Acknowledgments: This paper was written in grateful memory of our advisor and friend Thomas Beth.

Conflicts of Interest: The authors declare no conflict of interest.

References

1. Anshel, I.; Anshel, M.; Goldfeld, D. An Algebraic Method for Public-Key Cryptography. *Math. Res. Lett.* **1999**, *6*, 287–291. [CrossRef]
2. Ko, K.H.; Lee, S.J.; Cheon, J.H.; Han, J.W.; Kang, J.S.; Park, C. New Public-Key Cryptosystem Using Braid Groups. In Proceedings of the Advances in Cryptology—CRYPTO 2000, Santa Barbara, CA, USA, 20–24 August 2000; pp. 166–183.
3. Anshel, I.; Anshel, M.; Fisher, B.; Goldfeld, D. New Key Agreement Protocols in Braid Group Cryptography. In Proceedings of the Topics in Cryptology—CT-RSA 2001, San Francisco, CA, USA, 8–12 April 2001; pp. 13–27.
4. Grigoriev, D.; Ponomarenko, I. Constructions in public-key cryptography over matrix groups. In *Contemporary Mathematics: Algebraic Methods in Cryptography*; American Mathematical Society: Providence, RI, USA, 2006; Volume 418, pp. 103–119.
5. Lee, H.K.; Lee, H.S.; Lee, Y.R. An Authenticated Group Key Agreement Protocol on Braid groups. Cryptology ePrint Archive: Report 2003/018. 2003. Available online: http://eprint.iacr.org/2003/018 (accessed on 1 December 2019).
6. Shpilrain, V.; Ushakov, A. Thompson's Group and Public Key Cryptography. In Proceedings of the ACNS 2005—Third International Conference on Applied Cryptography and Network Security, New York, NY, USA, 7–10 June 2005; Volume 3531, pp. 151–163.

7. Shpilrain, V.; Zapata, G. Combinatorial group theory and public key cryptography. *Appl. Algebra Eng. Commun. Comput.* **2006**, *17*, 291–302. [CrossRef]
8. Shpilrain, V.; Ushakov, A. A new key exchange protocol based on the decomposition problem. In *Contemporary Mathematics: Algebraic Methods in Cryptography*; American Mathematical Society: Providence, RI, USA, 2006; Volume 418, pp. 161–167.
9. Anshel, I.; Anshel, M.; Goldfeld, D.; Lemieux, S. Key agreement, the Algebraic Eraser^{TM}, and lightweight cryptography. In *Contemporary Mathematics: Algebraic Methods in Cryptography*; American Mathematical Society: Providence, RI, USA, 2006; Volume 418, pp. 1–34. [CrossRef]
10. Anshel, I.; Atkins, D.; Goldfeld, D.; Gunnells, P.E. Ironwood Meta Key Agreement and Authentication Protocol. *arXiv* **2017**, arXiv:1702.02450.
11. Bellare, M.; Rogaway, P. Entitiy Authentication and Key Distribution. In Proceedings of the CRYPTO 1993—13th Annual International Cryptology Conference on Advances in Cryptology, Santa Barbara, CA, USA, 22–26 August 1993; Volume 773, pp. 232–249.
12. Bellare, M.; Canetti, R.; Krawczyk, H. A Modular Approach to the Design and Analysis of Authentication and Key Exchange Protocols. In Proceedings of the 30th Annual ACM Symposium on Theory of Computing STOC, Dallas, TX, USA, 24–26 May 1998; pp. 319–428.
13. Shoup, V. On Formal Models for Secure Key Exchange (Version 4). Revision of IBM Research Report RZ 3120 (April 1999). 1999. Available online: http://www.shoup.net/papers/skey.pdf (accessed on 1 December 2019).
14. Bellare, M.; Pointcheval, D.; Rogaway, P. Authenticated Key Exchange Secure Against Dictionary Attacks. In Proceedings of the EUROCRYPT 2000—Advances in Cryptology, Bruges, Belgium, 14–18 May 2000; Volume 1807, pp. 139–155.
15. Bresson, E.; Chevassut, O.; Pointcheval, D.; Quisquater, J.J. Provably Authenticated Group Diffie–Hellman Key Exchange. In *Proceedings of the 8th ACM Conference on Computer and Communications Security*; Samarati, P., Ed.; ACM Press: New York, NY, USA, 2001; pp. 255–264.
16. Canetti, R.; Krawczyk, H. Analysis of Key-Exchange Protocols and Their Use for Building Secure Channels. In *Advances in Cryptology—EUROCRYPT 2001*; Lecture Notes in Computer Science; Springer: Berlin/Heidelberg, Germany, 2001; Volume 2045, pp. 453–474.
17. Ben-Zvi, A.; Blackburn, S.R.; Tsaban, B. A Practical Cryptanalysis of the Algebraic Eraser. In *Advances in Cryptology—CRYPTO 2016 Proceedings, Part I*; Lecture Notes in Computer Science; Robshaw, M., Katz, J., Eds.; Springer: Berlin/Heidelberg, Germany, 2016; Volume 9814, pp. 179–189.
18. Cramer, R.; Shoup, V. Universal Hash Proofs and a Paradigm for Adaptive Chosen Ciphertext Secure Public-Key Encryption. In *Advances in Cryptology—EUROCRYPT 2002*; Lecture Notes in Computer Science; Knudsen, L., Ed.; Springer: Berlin/Heidelberg, Germany, 2002; Volume 2332, pp. 45–64.
19. González Vasco, M.I.; Martínez, C.; Steinwandt, R.; Villar, J.L. A new Cramer-Shoup like methodology for group based provably secure schemes. In *Proceedings of the 2nd Theory of Cryptography Conference (TCC 2005)*; Lecture Notes in Computer Science; Kilian, J., Ed.; Springer: Berlin/Heidelberg, Germany, 2005; Volume 3378, pp. 495–509.
20. Catalano, D.; Pointcheval, D.; Pornin, T. IPAKE: Isomorphisms for Password-based Authenticated Key Exchange. In *Advances in Cryptology—CRYPTO 2004*; Lecture Notes in Computer Science; Franklin, M.K., Ed.; Springer: Berlin/Heidelberg, Germany, 2004; Volume 3152, pp. 477–493.
21. Bohli, J.M.; Glas, B.; Steinwandt, R. Towards Provably Secure Group Key Agreement Building on Group Theory. In *Proceedings of VietCrypt 2006*; Lecture Notes in Computer Science; Springer: Berlin/Heidelberg, Germany, 2006; Volume 4341, pp. 322–336.
22. Abdalla, M.; Bohli, J.; González Vasco, M.I.; Steinwandt, R. (Password) Authenticated Key Establishment: From 2-Party to Group. In *Proceedings of the 4th Theory of Cryptography Conference TCC 2007*; Lecture Notes in Computer Science; Vadhan, S.P., Ed.; Springer: Berlin/Heidelberg, Germany, 2007; Volume 4392, pp. 499–514.
23. Burmester, M.; Desmedt, Y. A Secure and Efficient Conference Key Distribution System. In *Advances in Cryptology—EUROCRYPT'94*; Lecture Notes in Computer Science; Santis, A.D., Ed.; Springer: Berlin/Heidelberg, Germany, 1995; Volume 950, pp. 275–286.
24. Gennaro, R.; Lindell, Y. A Framework for Password-Based Authenticated Key Exchange. Cryptology ePrint Archive: Report 2003/032. 2003. Available online: http://eprint.iacr.org/2003/032 (accessed on 1 December 2019).

25. Gennaro, R.; Lindell, Y. A Framework for Password-Based Authenticated Key Exchange (Extended Abstract). In *Advances in Cryptology—EUROCRYPT 2003*; Lecture Notes in Computer Science; Biham, E., Ed.; Springer: Berlin/Heidelberg, Germany, 2003; Volume 2656, pp. 524–543.
26. Bohli, J.M.; González Vasco, M.I.; Steinwandt, R. Secure group key establishment revisited. *Int. J. Inf. Secur.* **2007**, *6*, 243–254. [CrossRef]
27. Bohli, J.M.; González Vasco, M.I.; Steinwandt, R. Password-authenticated group key establishment from smooth projective hash functions. *Int. J. Appl. Math. Comput. Sci.* **2019**, *29*, 797–815. Available online: http://eprint.iacr.org/2006/214 (accessed on 1 December, 2019). [CrossRef]
28. Katz, J.; Shin, J.S. Modeling insider attacks on group key-exchange protocols. In *Proceedings of the 12th ACM Conference on Computer and Communications Security (CCS 2005)*; Atluri, V., Meadows, C.A., Juels, A., Eds.; ACM: New York, NY, USA, 2005; pp. 180–189. Available online: http://eprint.iacr.org/2005/163 (accessed on 1 December, 2019).
29. Nam, J.; Paik, J.; Won, D. A security weakness in Abdalla et al.'s generic construction of a group key exchange protocol. *Inf. Sci.* **2011**, *181*, 234–238. [CrossRef]

© 2020 by the authors. Licensee MDPI, Basel, Switzerland. This article is an open access article distributed under the terms and conditions of the Creative Commons Attribution (CC BY) license (http://creativecommons.org/licenses/by/4.0/).

Article

Block Cipher in the Ideal Cipher Model: A Dedicated Permutation Modeled as a Black-Box Public Random Permutation

Yasir Nawaz and Lei Wang

Department of Computer Science and Engineering, Shanghai Jiao Tong University, Shanghai 200240, China; my_nawaz@sjtu.edu.cn (Y.N.); wanglei@cs.sjtu.edu.cn (L.W.)

Received: 3 November 2019; Accepted: 2 December 2019; Published: 5 December 2019

Abstract: Designing a secure construction has always been a fascinating area for the researchers in the field of symmetric key cryptography. This research aimed to make contributions to the design of secure block cipher in the ideal cipher model whose underlying primitive is a family of $n-bit$ to $n-bit$ random permutations indexed by *secret key*. Our target construction of a secure block ciphers denoted as $\mathbb{E}[s]$ is built on a simple *XOR* operation and two block cipher invocations, under the assumptions that the block cipher in use is a pseudorandom permutation. One out of these two block cipher invocations produce a subkey that is derived from the secret key. It has been accepted that at least two block cipher invocations with *XOR* operations are required to achieve beyond birthday bound security. In this paper, we investigated the $\mathbb{E}[s]$ instances with the advanced proof technique and efficient block cipher constructions that bypass the birthday-bound up to 2^n provable security was achieved. Our study provided new insights to the block cipher that is beyond birthday bound security.

Keywords: pseudorandom permutation; block cipher; ideal cipher model; beyond birthday bound; provable security

1. Introduction

A block cipher encryption design is called *beyond birthday bound (BBB)* secure if the proven upper bound on the adversarial advantage is meaningful even if an adversary can process more than $2^{n/2}$ data blocks, where n is the size of the block of a block cipher. The first time, Iwata proposed a *BBB* encryption mode cipher-based encryption (*CENC*) [1]. This was nonce based construction providing a solution through the invocation of more than one block cipher and simple *XOR* operation and achieved $2^{2n/3}$ security against all nonce respecting adversaries. Later on, Iwata proved *CENC* construction based on mirror theory technique [2], and achieved optimal security [3]. Bhattacharya and Nandi also gave the *BBB* security of *CENC* by analyzing the security bound of variable output length using the chi-squared method.

1.1. Pseudorandom Permutation and Pseudorandom Function with BBB

The conventional approach for designing the cryptography primitives based on symmetric cipher is to behave as a perfectly random function. The vast majority, in this case, is an encryption scheme [4], *MAC* encryption schemes [5,6], and authenticated encryption schemes [7], following this paradigm via pseudorandom functions (*PRF*). Patarin suggested the construction of permutation sum and proved that a variant of single permutation indistinguishable from a random function up to *BBB* [8]. In 2003, Patarin gave the result $2^{2n/3}$ security [9], like so, in 2005, achieved up to this security bound [10,11]. However, the *PRF* provides a solution for increasing the use of cryptography in a real-world application. The pseudorandom permutation (*PRP*) is the leading building block of the cryptographic design in spite of *PRF* [12–15]. If a block cipher is directly implemented as a *PRF*, which will have provable security limit birthday bound with a large block, this is often acceptable. But it is not acceptable in practice with a lightweight block cipher, which has relatively small block sizes. The *PRF* can be replaced by a *PRP* up to birthday bound queries [16–19]. Moreover, if the block size of a block cipher is large enough, then the security loss is sometimes acceptable. Whatever, there are many scenarios, such as lightweight applications, whose numbers have grown tremendously before some years that require higher security bound [20–26]. In recent years, various constructions have been proposed that achieve *BBB* security against more than $2^{n/2}$ queries. We could categorize these constructions into *XOR permutations based* and *truncation based*. The *XOR permutations* is popular for *BBB* construction by taking the *XOR* of more than one independent *PRP* [20].

$$XOR_{E_{k_1}, E_{k_2}}(x) = E_{k_1}(x) \oplus E_{k_2}(x)$$

This construction was analyzed by Lucks [21]. The single key variant of this construction provides the security up to $2^{2n/3}$ queries [27]. After that, Patarin revised this construction and improved the security bound up to $2^{n/67}$ [28]. Later on, the results were generated by more than two independent *PRP* with *XOR* operation [29]. Dai et al. [30] using the chi-squared method verified the $n - bit$ security of *XOR* construction, but the original proof was provided by Bhattacharya and Nandi [31]. The *XOR* construction is acceptable for encryption, but it is not usable for authentication, because domain size is required to extend. This can be solved through hashing the message, but the *XOR* construction needs some precise combination with a double block hash function [32–34]. The truncation based solution was presented by Hall et al. [17]. Later on, it was proved that truncating $n - bit$ permutation has security bound up to $2^{2n/3}$ queries [35]. Stam also derived these results in a non-cryptographic context [27]. Recently, another construction was proposed, which is known as *Encrypted Davies Meyer* (*EDM*) introduced by Cogliati and Seurin [36].

$$EDM_{E_{k_1}, E_{k_2}}(x) = E_{k_2}(E_{k_1}(x) \oplus x)$$

There are two independent permutations and it behaves like random function up to $q^3/2^{2n}$ [36]. Afterward, Dai et al. [30] achieved $q^4/2^{3n}$ using the chi-squared method. Now, a novel construction *EDMD* improved the security up to $2^n/67n$ by using mirror theory technique, which has almost an optimal security [37].

$$EDMD_{E_{k_1}, E_{k_2}}(x) = E_{k_2}(E_{k_1}(x)) \oplus E_{k_1}(x)$$

Two independent keys are required for *EDMD*. The single key setting is significant for higher security bound and efficient construction, which was also performed in our construction. Anyways, this construction secures up to $q/2^{2n/3}$. Cogliati and Seurin also extended the *EDMD* construction called *encrypted Wegman carter with davies meyer* (*EWCDM*), which is nonce based *BBB* secure.

$$EWCDM_{E_{k_1}, E_{k_2}, H_{k_h}}(N, M) = E_{k_2}(E_{k_1}(N) \oplus N \oplus H_K(M))$$

where, H_K is a universal hash function, N denote the *nonce*, and M denote the *message*, which has an arbitrary length. The *EWCDM* achieved *BBB* up to $2^{2n/3}$ *MAC* queries when it has nonce respecting setting. The use of internal state values of *EWCDM* construction makes their security analysis formally inapplicable [37]. Mennink presented the rationale relying on the *EWCDM* function, and simplified versions of the conversion method applied to the advanced encryption scheme (*AES*) [38]. The main proposal of *AES-PRF*, the *AES* with a feed forward of the middle state, achieved almost no optimal security. This construction was applied to *GCM* and *GCM-SIV*, and how it entails the significant security improvements was discussed. A little while back, Mennink presented a heuristic study to build *BBB* secure from public random permutation, showing that a single permutation call could not be secured *BBB* [39].

The above discussion shows that what to be tackled in *PRF* for *BBB* and where the goal is to build *PRF*, so that it is indistinguishable from a truly random function. However, our study aimed to build block cipher in the ideal cipher model, under the assumption that the block cipher is a *PRP* out of

PRF, achieving *full security*. Moreover, the *sum of even mansour* (*SoEM*) construction achieves *BBB* up to $2^{2n/3}$, that is built from two randomly drawn keys and two independent permutations; if either keys or permutations are identical, then there is a birthday bound attack.

1.2. Our Construction

In this paper, we focused on a block cipher design based on a single key, which achieved *BBB* up to 2^n security. The main motivation is by the scenarios where the block cipher only has block size of $32 - bit$, $48 - bit$, and $64 - bit$ [40]. The target construction of block cipher depicted in Figure 1, defined as $\mathbb{E}[s] : K \times P \to P$, consists of two block cipher invocations and additional simple *XOR* operation. Furthermore, a heuristic approach is carried out to examine the instances of $\mathbb{E}[s]$ and, at last, $E1 - E32$ efficient construction is successfully found. In detail, the first invoke of block cipher produces a subkey y from the secret key k such that $y = E(k,0)$, $y = E(0,k)$, and $y = E(k,k)$. The second invoke of a block cipher encrypt and decrypt the plaintext p and ciphertext c, respectively, with a key k or $k \oplus y$. However, we stress that the first block cipher invocation is precomputing and storing the subkey y. Thus, our design only requires one invocation of a block cipher for encryption and decryption when the subkey y is precomputed and stored. We have designed this construction in the ideal cipher model that has the main advantage of provable security up to 2^n. The previously available block cipher has maximum provable security up to $2^{2n/3}$. From the efficiency point of view, previous constructions required more than one key, $s > 2$ block cipher invocations [20,36], and universal hash function invocations; in the absence of these, their efficiency needed to be increased. The minimum number of block cipher invocation with a single key is good for efficiency. Our design requires just a single secret key and one block cipher invocation for encryption and decryption when the subkey is precomputed and stored.

2. Preliminaries

2.1. Notations

The $\{0,1\}^n$ denotes the set of bit strings of length n. We denote the bitwise addition $a \oplus b$, where $a, b \in \{0,1\}^n$. The $Y \leftarrow Z$ is the assignment of Z to the variable Y. The $x \xleftarrow{\$} X$ denotes the uniform random selection of x from X. The $|X|$ denotes the number of elements in X. Let $a \in \{0, 1\}$ and $b \in \{0, 1\}$, $a.b$ denotes the multiplication of a and b, if $a = 1$, then it is equal to b, and if $a = 0$, then $a.b$ equals to 0. The block cipher denotes as $E : K \times P \to P$, where P is a plaintext/message space, K is the key space. Throughout the paper, we have fixed $K = P = \{0,1\}^n$. Let $E(k, \cdot)$ and $E^{-1}(k, \cdot)$ denote the encryption and decryption, respectively, with a secret key $k \in K$. Let $E^{\pm}(k, \cdot)$ involves $E(k, \cdot)$ and $E^{-1}(k, \cdot)$. Sometimes, we denote $E(k, \cdot)$ as $E_k(\cdot)$, $E^{-1}(k, \cdot)$ as $E_k^{-1}(\cdot)$, and $E^{\pm}(k, \cdot)$ as $E_k(\cdot)$ and $E_k^{-1}(\cdot)$, respectively. The (u, w) are the input and output tuple of E such that $w = E(u)$. The input-output tuple of E_k is denoted as (p, c) such that $E_k(p) = c$. Let $Perm(n)$ denote the set of all permutations on $\{0,1\}^n$. The function π is said to be an ideal cipher model if randomly selected that is $\pi \xleftarrow{R} Perm(n)$. Similarly, we define these notations $\pi(\cdot, \cdot)$, $\pi^{-1}(\cdot, \cdot)$, and $\pi^{\pm}(\cdot, \cdot)$, respectively.

2.2. Security Definition

A computationally unbounded distinguisher D is an algorithm that has adaptive access to an oracle and outputs a bit 0 or 1. Let the two oracles O_1 and O_2 have the same interface, we can get the distinguishing advantage of D as follows.

$$Adv(D) = \Pr[D^{O_1} \Rightarrow 1] - \Pr\left[\Pr[D^{O_2} \Rightarrow 1\right]$$

A block cipher with a key space K and message space P is a mapping $E : K \times P \to P$ such that for all key $k \in K$. The $E(K, P)$ is a permutation over P. We denote $E_k(P)$ for $E(K, P)$. The distinguisher D is having query access to $(O_1, E^{\pm})$: O_1 is either $E_k^{\pm}(\cdot, \cdot)$ with $k \xleftarrow{\$} K$ or $\pi \xleftarrow{\$} Perm$. The $E^{\pm}$ is an underlying block cipher. The advantage of distinguisher D in distinguishing E and π is defined as.

$$Adv_E^{prp}(D) = \left| \Pr[D^{E_k^\pm(\cdot,\cdot), E^\pm(\cdot,\cdot)} \Rightarrow 1] - \Pr[D^{\pi^\pm(\cdot,\cdot), E^\pm(\cdot,\cdot)} \Rightarrow 1] \right|$$

Throughout this paper, we considered information as theoretical with computationally unbounded distinguishers D sorely limited by the number of queries to the oracle. Overall, maximum is taken by distinguisher D that makes at most q queries to its oracles.

$$Adv_E^{prp}(q) = max_D \{ Adv_E^{prp}(D) \}$$

2.3. H-Coefficient Technique

Central to our proof is a *H-Coefficient technique* presented by Patarin [8,41]. As mentioned above, we considered the information as theoretical, with computationally unbounded distinguisher D. Thus, we always assumed that distinguisher D is deterministic without the loss of generality. Let distinguisher D interacts with O_1 and O_2. The interaction of D with its oracles are recorded in a view v. The X_{O_2} is the probability distribution of v when distinguisher D interacts with O_2. The V is the set of all attainable views v when D interacting with O_2, which is $V = \left\{ v \big| \Pr[X_{O_2} = v] > 0 \right\}$. The H-Coefficient technique states as follows:

Let $0 \leq \varepsilon \leq 1$. Consider a partition $V = V_{good} \cup V_{bad}$ set of attainable view such that:

1. $\Pr[X_{O_2} \in V_{bad}]$
2. for all $v \in V_{good}$, $\frac{\Pr[X_{O_1} = v]}{\Pr[X_{O_2} = v]} \geq 1 - \varepsilon$

Then, the distinguishing advantage satisfies

$$Adv(D) \leq \Pr[X_{O_2} \in V_{bad}] + \varepsilon$$

The core idea of the H-coefficient technique is: a large number of views are almost equally likely in both oracles (real worlds and the ideal world), and the odd ones occur with a small probability. Note that the partitioning of V into *bad* and *good* views is directly reflected in the terms $\Pr[X_{O_2} \in V_{bad}]$ and ε in the bound: if V_{good} is too large, ε will become large, whereas if V_{bad} is too large, $\Pr[X_{O_2} \in V_{bad}]$ will become large.

3. Construction Limitations

In this section, we will discuss the construction limitations of secure block cipher in the ideal cipher model, which is built on dedicated block cipher invocations and simple XOR operation. The XOR operation has efficiency benefits. The target construction is denoted as $\mathbb{E}[s]$ and is built on s block cipher invocations. Let E denote the underlying block cipher with $n-bit$ block size and $n-bit$. key size. Let p, c, and k denote the plaintext, ciphertext, and key, respectively, where all have $n-bit$ size. Let $a_{i,j}$ and $b_{i,j}$ be one bit variable of being 0 or 1, where $1 \leq i \leq s + 1$ and $1 \leq j \leq i + 2$. The encryption of $\mathbb{E}[s]$ is shown in Algorithm 1. The target construction $\mathbb{E}[s]$ is depicted in Figure 1. In detail, this is a graphical view from which we would acquire the resultant block cipher construction. Moreover, all the s block cipher invocations are involved in the computation of the ciphertext c. The ciphertext c must be invertible and efficiently decrypted from plaintext p and key k. There are some limitations for $\mathbb{E}[s]$ to achieve our goal:

- The plaintext p should be involved in exactly one XOR operation. The p involves in XOR operation, which gives x_i and corresponding y_i. So, both outputs (x_i and y_i) are called *plaintext dependent variable*. On the other side, if a variable y_i is used to compute another variable x_j, which depends on y_i, then x_j and corresponding y_j would also be plaintext dependent variable. So, we cannot use plaintext dependent variable to produce any key or subkey, otherwise, constructions will not be efficient.

There should be at most one plaintext dependent variable produced from the *XOR* operation. Otherwise, the decryption process cannot efficiently decrypt because there is more than one variable.

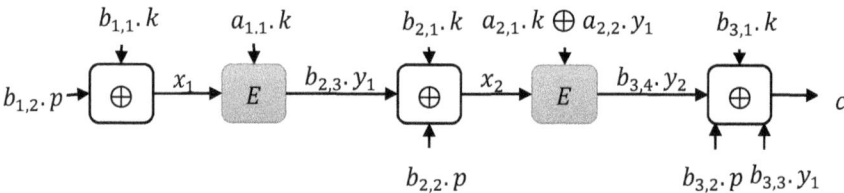

Figure 1. $\mathbb{E}[s]$: Target Construction.

If we summarize and satisfy the above limitations, then $\mathbb{E}[s]$ can be an efficient block cipher construction. Moreover, an additional condition is also necessary for efficiency and security. Our first goal is to achieve full (2^n) provable security. The target construction is important to achieve the goal. Nowadays, *AES* and *SIMON* block cipher is utilized in various applications of different block sizes, such as $128 - bit$ and $64 - bit$. In some environments, the block size of lightweight block ciphers can be even shorter. Thus, block cipher construction with a simply birthday bound security may not be suitable for various applications. Therefore, another construction which provide higher security is definitely necessary. Particularly, for application design, a block cipher with full security is surely an interesting research topic. Our second goal is the efficiency, we invoke two block cipher because minimum number of block cipher invocation led to concern about high efficiency. It is well known that block cipher invocations are much more time consuming than *XOR* operation. So, the efficiency reduces due to a number of block cipher invocation. But, besides this, we aimed to achieve perfect efficiency under the condition of no security sacrifices, i.e., eliminating the unnecessary input variables. In fact, this is also a reason in our target construction having simple *XOR* operation and only necessary input variables. Algorithm 1 is shown as follow:

Algorithm 1 $\mathbb{E}[s](\cdot,\cdot)$

---

input: k, p, $E(\cdot,\cdot)$, vaiables $a_{i,j}$ and $b_{i,j}$
Output: *ciphertext* $x_1 = a_{1,1}.k$, $b_{1,1}.k \oplus b_{1,2}.p$

1. $x_1 = a_{1,1}.k$, $b_{1,1}.k \oplus b_{1,2}.p$
2. for $i = 1$ to $s - 1$, do
3. $y_i = E(a_{1,1}.k, x_i)$
4. $x_{i+1} = a_{i\oplus 1,1}.k \oplus \sum_{j=2}^{i+1} a_{i\oplus 1,j}.y_{j-1}, b_{i\oplus 1,1}.k \oplus b_{i\oplus 1,2}.p \oplus \sum_{j=3}^{i+2} b_{i\oplus 1,j}.y_{j-2}$
5. end for
6. $y_s = E(k_s, x_s)$
7. $c = b_{s\oplus 1,1}.k \oplus b_{s\oplus 1,2}.k \oplus \sum_{j=3}^{s+2} b_{s\oplus 1,j}.y_{j-2}$
8. return ciphertext c

In order to achieve the above goals among the instances of target construction, we adopted a heuristic approach. For the instances of $\mathbb{E}[s]$, we invoked only two block cipher to achieve 2^n provable security because $s = 1$ for instances of $\mathbb{E}[s]$ had most $2^{n/2}$ security. Thus, at least two block cipher invocations are required to bypass the birthday bound barrier.

We continued to examine the instances of $\mathbb{E}[2]$ and would not analyze the $\mathbb{E}[s > 2]$ instances unless investigated all the instances of $\mathbb{E}[2]$ and none of them achieve 2^n security. In fact, if some instances of $\mathbb{E}[2]$ achieves 2^n security, then there is no need to examine the other instances of $\mathbb{E}[2]$. To follow the above strategy, we analyzed the target construction $\mathbb{E}[s]$ and found 32 instances with 2^n provable security.

3.1. $\mathbb{E}[2]$ Instances

According to the previous discussion, the plaintext p should be involved in exactly one XOR operation. There should be, at most, one plaintext dependent variable produced from the XOR operation. Otherwise, the decryption process cannot efficiently decrypt because there exists more than one variable. The plaintext dependent variable cannot be used to produce any key-value; otherwise, constructions will not be efficient. Following this strategy, we divided $\mathbb{E}[2]$ instances into three types on the basis of when plaintext p is XOR to compute x_i and c, respectively.

- Type 1 instances: when p is XOR to compute x_1
- Type 2 instances: when p is XOR to compute x_2
- Type 3 instances: when p is XOR to compute c

3.1.1. Type 1 Instances

According to the above limitation, the plaintext dependent variables cannot be used to produce key value, so, $a_{2,2} = 0$. The plaintext p should be involved in exactly one XOR operation, so, $b_{2,2} = 0$ and $b_{3,2} = 0$. We set $b_{2,3} = 1$, which is the first block cipher invocation, and set $b_{3,4} = 1$, which is second block cipher invocation. If $b_{2,3} = 0$, it means two block ciphers' invocations are parallel, and these instances are involved in type 2. It also shows that x_2 and y_2 are plaintext variables. Then, we set $b_{3,3} = 0$ because y_2 is already used as a plaintext dependent variable. All of these simplified constructions of type 1 are shown in Figure 2. We examined the instances of type 1, and ciphertext is computed as follows.
$$c = E(a_{2,1}.k, x_2) \oplus b_{3,1}.k$$

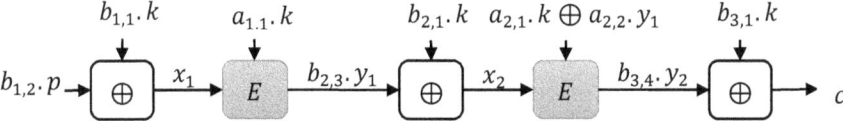

Figure 2. $\mathbb{E}[2]$: Type 1 Construction.

Instances with one block cipher Invocation of type 1.

We would show that any instance that makes only one block cipher invocation of type 1 construction could not achieve BBB security. Let $E : \{0,1\}^n \times \{0,1\}^n \to \{0,1\}^n$ be a block cipher, shown in Figure 3. We showed that there exists a distinguisher D that can distinguish any such block cipher from random permutation using at most $2^{n/2}$ queries.

- When $a_{1,1} = 0$ and $b_{1,1} = 1$.

In this case, we can see the input or output of E is not related to p or c. When $b_{1,2} = 0$, then distinguisher D selects arbitrary p and p' to get c and c'. If the event $c = c'$ occurs, then output is 1; otherwise, it is 0. The success probability of D is 1 when interacts with $1 - 2^{-n}$. The results are similar for $b_{2,3} = 0$.

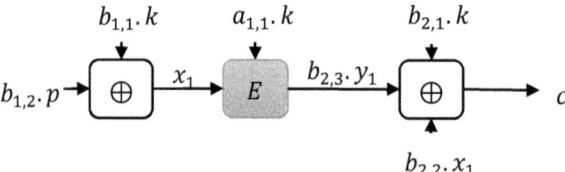

Figure 3. Type 1: One Block cipher invocation.

- When $a_{1,1} = 0$ and $b_{1,1} = 0$.

In this case, we can see the input or output of E is independent of the key. When $b_{1,2} = 1$, the distinguisher D selects arbitrary x_1 and x_1' to get y_1 and y_1'; then, it puts $p = b_{1,2}^{-1} x_1$ and $p' = b_{1,2}^{-1} x_1'$ to get c and c'. If the event occurs, then output is 1, otherwise 0.

$$Event = \begin{cases} c \oplus c' = b_{2,3}.y_1 \oplus b_{2,3}.\ y_1' \ if \ b_{2,2}.x_1 = 0 \\ c \oplus c' = b_{2,3}.y_1 \oplus b_{2,3}.\ y_1' \oplus x_1 \oplus x_1' \ if \ b_{2,2}.x_1 \neq 0 \end{cases}$$

The success probability of D is 1 when interacts with $1 - 2^{-n}$. Similar is the case for $b_{2,1} = 0$.

- When $b_{2,2} = 0$.

In this case, there exists a distinguisher D, distinguishing the real world oracle $\left(E_k^{\pm}, E^{\pm}\right)$ from the ideal world oracle $(\pi^{\pm}, E^{\pm})$ with some probability. The distinguisher D makes $2^{n/2}$ queries and operates as follows. For $j = 1, \ldots, 2^{n/2}$, the distinguisher D selects arbitrary $p^{(j)}$ to get $c^{(j)}$. If $c^{(j)} \neq c^{(j')}$ for all queries and its indices $j \neq j'$, then output 1, otherwise output 0.

At the end of type 1 instances, we can conclude that the plaintext added in the first XOR operation and the output value after the first invocation of block cipher are included in second block cipher invocation as a key that is a plaintext dependent variable, so the advantage of the adversary is at most around birthday bound.

3.1.2. Type 2 Instances

Following the construction limitations, set $b_{3,5} = 1$. The plaintext p should be involved in exactly one XOR operation, so, $b_{1,2} = 0$ and $b_{3,2} = 0$. We set $b_{2,3} = 1$, that is, the first block cipher invocation, and thus, we set $b_{3,4} = 1$, that is, second block cipher invocation. It also shows that x_1 and y_1 are not plaintext dependent variables. All of these simplified constructions of type 1 are depicted in Figure 4. Here, we examined the type 2 instances. For these instances, we computed ciphertext as follows.

$$c = E(a_{2,1}.k \oplus b_{3,3}.y_1,\ x_2) \oplus b_{3,1}.k \oplus b_{3,3}.y_1$$

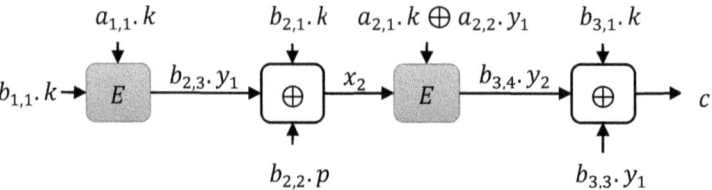

Figure 4. $\mathbb{E}[2]$: Type 2 Construction.

The first block cipher invocation is $y_1 = E(a_{1,1}.k, b_{1,1}.k)$. Throughout all the instances of type 2, we call y_1 as a subkey that is obtained from the secret key k for those instances with $(a_{1,1}, b_{1,1}) \neq (0,0)$.

However, the computation from p to x_2 is $x_2 = p \oplus b_{2,1}.k \oplus b_{2,3}.y_1$, and $\Delta x_2 = \Delta p$ always holds and $\Delta y_2 = \Delta c$, respectively. Moreover, for any plaintext and ciphertext pair (p,c) and (p',c'), the adversary knows the internal variable differences Δx_2 and Δy_2. Therefore, according to the above constraint, we can find some conditions on the type 2 instances to achieve BBB.

- When $(a_{1,1}, b_{1,1}) \neq (0,0)$.

  If $(a_{1,1}, b_{1,1}) = (0,0)$, then it means $y_1 = E(0,0)$. Adversary makes a query $(0,0)$ to $E(\cdot, \cdot)$ to get y_1, and the first block cipher invocation kicks off. Then, the instances are based on only a single block cipher invocation in the adversary view. As we discussed in the previous sections, when $s < 2$, the construction achieves security up to birthday bound.

- When $(a_{2,1}, a_{2,2}) \neq (0,0)$.

  If $(a_{2,1}, a_{2,2}) = (0,0)$, then adversary regards $b_{2,1}.k \oplus b_{2,3}.y_1$ and $b_{3,1}.k \oplus b_{3,3}.y_1$. So, the instance gives essentially one step of [42].

- When $(b_{2,1}, b_{2,3}) \neq (0,0)$.

  If $(b_{2,1}, b_{2,3}) = (0,0)$, then $p = x_2$, i.e., the adversary knows and can control the x_2 value. A distinguisher D is launched and fixes two distinct p and p'. The distinguisher D queries to $\mathbb{E}[2]_k(\cdot,\cdot)$ and gets ciphertext c and c' and stores $(c \oplus c')$, respectively. The D makes a query for $E(\cdot, \cdot)$ and receives ω and $\acute{\omega}$, respectively, and matches $\omega \oplus \acute{\omega}$ to stored $c \oplus c'$. The distinguisher D recovers $a_{2,1}.k \oplus a_{2,2}.y_1$. For any plaintext-ciphertext pair (p,c) and (p',c'), the distinguisher D can compute z (such that $a_{2,1}.k \oplus a_{2,2}.y_1 = z$) and z' and query (z,p) and (z',p') to $E(\cdot, \cdot)$, recovering y_2 and y'_2, respectively. So, the output of distinguisher D is 1 if $c \oplus c' = y_2 \oplus y'_2$, otherwise, compute 0. When interacting with $\mathbb{E}[2]$, then the output of distinguisher D is 1 until it recovers $a_{2,1}.k \oplus a_{2,2}.y_1$. Thus, the success probability is $1 - (1 - 2^{-n})^{2^n}$.

- When $(b_{3,1}, b_{3,3}) \neq (0,0)$.

  This has a similar analysis which is presented above, where the adversary knows and has control over the value of y_2 and he fixes the ciphertext c and c' and queries to $\mathbb{E}[2]_k^{-1}(\cdot,\cdot)$.

- When $(b_{2,1}, b_{2,3}) \neq (a_{2,1}, a_{2,2})$.

  If $(b_{2,1}, b_{2,3}) = (a_{2,1}, a_{2,2})$, it has $(b_{2,1}.k \oplus b_{2,3}.y_1) = (a_{2,1}.k \oplus a_{2,2}.y_1)$, which is denoted by g and $x_2 \oplus z_2 = g \oplus p \oplus g = p$. Thus, the adversary knows and can control $x_2 \oplus z$. A distinguisher D is launched and gives queries to $\mathbb{E}[2]_k(\cdot, \cdot)$ and receives c and c' and stores $(c \oplus c')$, respectively. Moreover, D sends distinct queries to $E(\cdot, \cdot)$ and receives ω and $\acute{\omega}$, respectively, and stores $(\omega \oplus \acute{\omega})$. Then, he matches $(\omega \oplus \acute{\omega})$ and $(c \oplus c')$. The D can compute x_2 and z for any plaintext-ciphertext and receive y_2 from $E(\cdot, \cdot)$. Moreover, the distinguisher D just needs to make some extra queries. Thus, the success probability is trivially $1 - (1 - 2^{-n})^{2^n}$.

- When $(b_{3,1}, b_{3,3}) \neq (a_{2,1}, a_{2,2})$.

  This is also having a similar analysis as shown above.

  Putting all the above properties of type 2 instances together, we got 32 instances, denoted by E1, E2,..., E32 and depicted in Figure 5. We investigated these constructions and found 2^n provable security. We used the H-Coefficient technique for proof, which is discussed in Section 4.

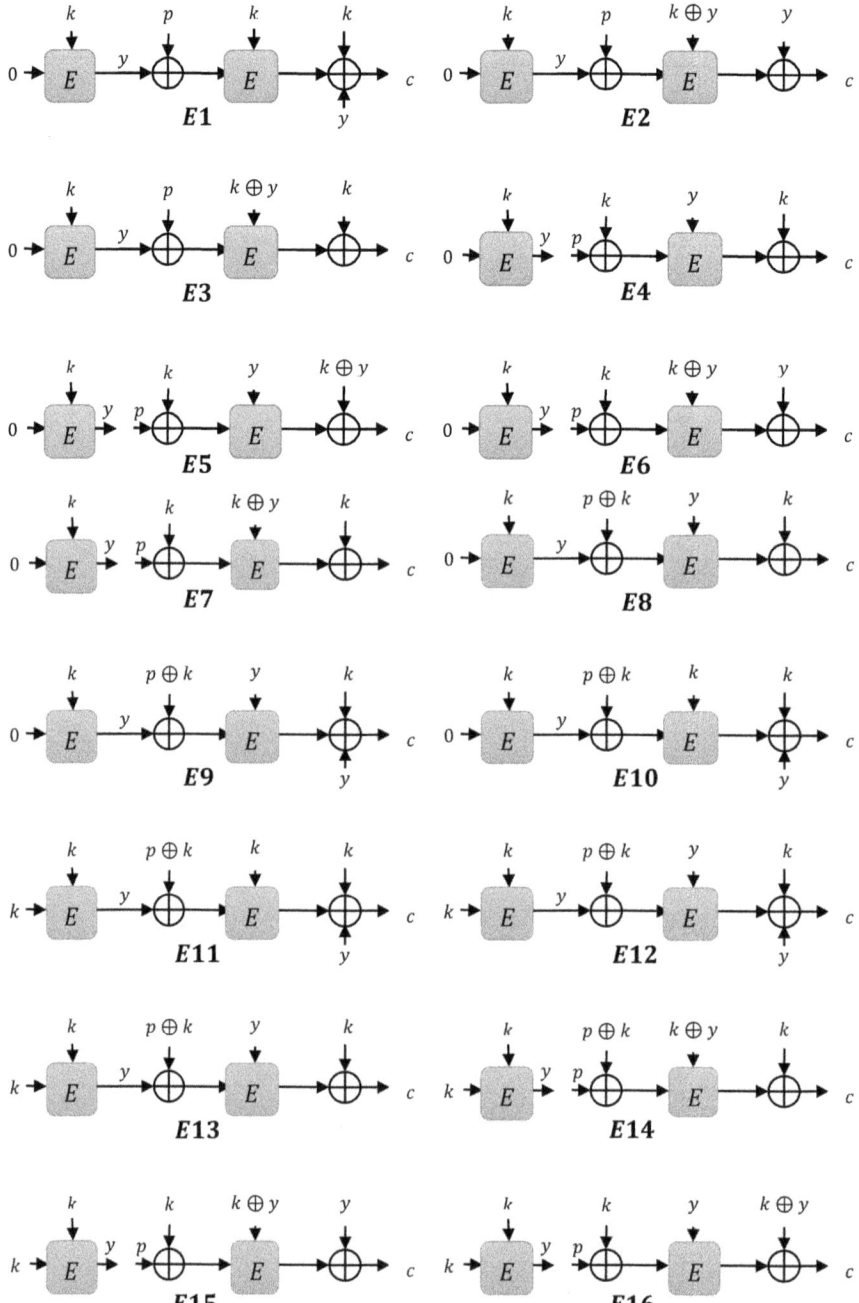

Figure 5. *Cont.*

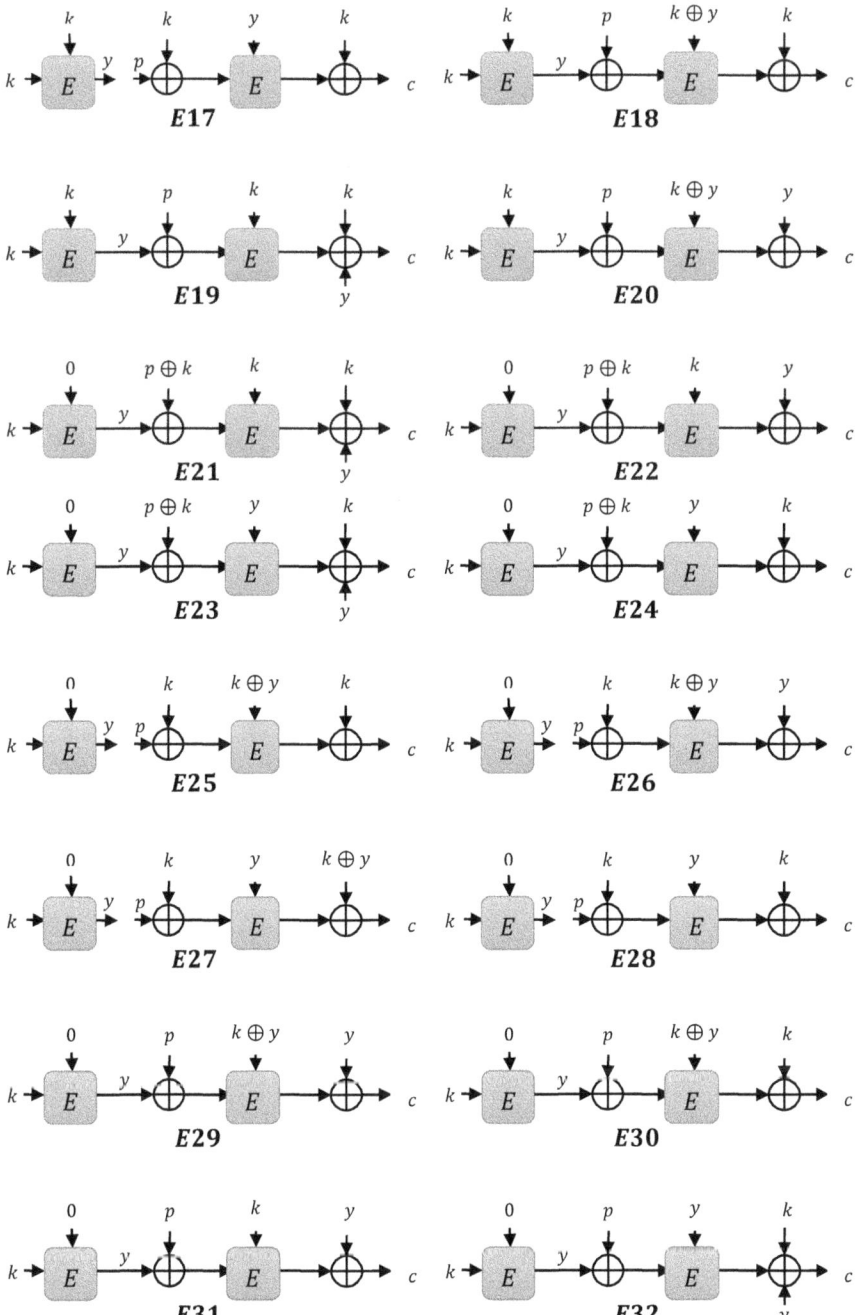

Figure 5. The $E1, E2, \ldots,$ and $E32$ efficient construction: the internal variable y is referred to as a subkey for these constructions.

3.1.3. Type 3 Instances

When p is XOR to compute c, then $b_{3,2}.p = 1$, $b_{1,2}.p = 0$, and $b_{2,2}.p = 0$. The constructions of type 3 are depicted in Figure 6. In this construction, it could be seen that p and c are linearly related, and distinguisher D can distinguish by only two queries to $\mathbb{E}[2]_k(\cdot,\cdot)$ with distinct plaintext p and $p \oplus \Delta$, verifying $\Delta c = \Delta$. Hence, the discussion of type 3 instances is omitted here.

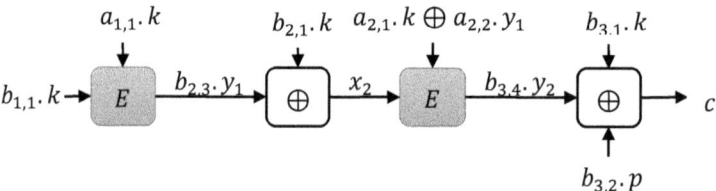

Figure 6. $\mathbb{E}[2]$: Type 3 Construction.

4. Security Proof

Let $E1, E2, \ldots, E32$ is any instance, and E is an underlying block cipher. Let there be any distinguisher D that has access to oracles O_1 and O_2, either $E_k^{\pm}(\cdot,\cdot), E^{\pm}(\cdot,\cdot)$ with $k \xleftarrow{\$} K$ or $\pi^{\pm}(\cdot,\cdot), E^{\pm}(\cdot,\cdot)$. The distinguisher D is computationally unbounded and deterministic, making q queries when interacting with O_1 and O_2. We defined distinguisher queries to O_1 and O_2 as q_1 and q_2, respectively: $q = q_1 + q_2$ and do not contain duplicate queries. When distinguisher D interacts with O_1 and O_2, the queries response are $v_1 = \{(p_1, c_1), \ldots, (p_{q_1}, c_{q_1})\}$ and $v_2 = \{(u_1, w_1), \ldots, (u_{q_2}, w_{q_2})\}$, respectively. The v is the view denoting the transcripts, and in the end, the distinguisher D obtains a view $v = (v_1, v_2)$. The distinguisher D, based on the v, computes its decision bit. Accordingly, the decision *bit* probability distribution of distinguisher D relies on the probability distribution of v. The X and Y are the probability distribution on v when interacts with $(E_k^{\pm}(\cdot,\cdot), E^{\pm}(\cdot,\cdot))$ and $(\pi^{\pm}(\cdot,\cdot), E^{\pm}(\cdot,\cdot))$, respectively. We used V as an attainable view when D interacts with O_1, which is $V = \{v | \Pr[Y = v] > 0\}$ and $V = V_{good} \cup V_{bad}$. The main goal of the proof is to disclose the subkey y and secret key k after interacting with O_1 and O_2. In $(\pi^{\pm}(\cdot,\cdot), E^{\pm}(\cdot,\cdot))$ as (O_1, O_2), we chose $k \xleftarrow{\$} K$ and got corresponding subkey y by querying $E^{\pm}$. The distinguisher D can easily derive query response (u, w) of $E^{\pm}(\cdot,\cdot)$ invocations for each query response (p_i, c_i) in view v_1. The query responses of a block cipher E for each view $v = (v_1, v_2) \in V$ is divided into three tables. The first one consists of a single query response of block cipher E: $T^1 = \{(u_1^1, w_1^1 = y)\}$. The second table consists of the other queries' responses of block cipher E derived from v_1: $T^2 = \{(u_1^2, w_1^2), \ldots, u_{q_2}^2, w_{q_2}^2\}$. The last table consists of all queries' responses from v_2: T^3: $T^3 = \{(u_1^3, w_1^3), \ldots, u_{q_2}^3, w_{q_2}^3\}$.

4.1. Bad Events

$v \in V_{bad}$ if there are following queries: $T^1 = \{(u_1^1, w_1^1 = y)\}$, $T^2 = \{(u_1^2, w_1^2), \ldots, u_{q_2}^2, w_{q_2}^2\}$, and $T^3 = \{(u_1^3, w_1^3), \ldots, u_{q_2}^3, w_{q_2}^3\}$ such that the following condition holds: there exists (u_j^i, w_j^i) in table T^i and $(u_{j'}^{i'}, w_{j'}^{i'})$ in table $T^{i'}$ such that $(u_j^i, w_j^i) = (u_{j'}^{i'}, w_{j'}^{i'})$ where $i \neq i'$, then v causes bad event.

4.2. $\Pr[Y \in V_{bad}]$

According to our construction, we gave here the exact definition of V_{bad}, which also ensures the V_{good}. The V_{good} does not cause bad event. Here, we defined the V_{bad} of $E1$ only due to the limited space. At least, one event defines the V_{bad} if it exists.

(a) $(p_i, c_i) \in v_1$ such that $p_i = y$;
(b) $(p_i, c_i) \in v_1$ such that $c_i = k$;
(c) $(p_i, c_i) \in v_1$ and $(u_j, w_j) \in v_2$ such that $(u_j = p_i \oplus y)$
(d) $(p_i, c_i) \in v_1$ and $(u_j, w_j) \in v_2$ such that $(w_j = c_i \oplus y \oplus k)$

The subkey y and secret k are uniformly selected at random from a set of size of at least $2^n - q - 1$. We get

$$\Pr[(a)] \leq q/2^n - q - 1;$$
$$\Pr[(b)] \leq q/2^n - q - 1;$$
$$\Pr[(c)] \leq q/2^n - q - 1;$$
$$\Pr[(d)] \leq q/2^n - q - 1;$$

Thus, we get

$$\Pr[Y \in V_{bad}] \leq \Pr[(a)] + \Pr[(b)] + \Pr[(c)] + \Pr[(d)]$$

Let $q < 2^{n-1}$ and using above values, we get

$$\Pr[Y \in V_{bad}] \leq \frac{4q}{2^{n-1}}$$

4.3. Ratio for V_{good}

First of all, $\Pr[X = v]$. The X is a random variable that is defined on the probability space of all possible underlying block cipher E and all possible secret key k. The probability space of X is denoted as all_X. Correspondingly, the $|all_X|$ is equal to $2^n (2^n!)^{2^n}$. In all_X, an element π getting along with v is taken, if π gives exactly the same responses for all queries. The $comp_X(v)$ is defined as all the elements in all_X compatible with v.

$$\Pr[X = v] = \frac{|comp_X(v)|}{all_X}$$

Similarly, Y is defined on the probability space of $E1$, underlying block cipher E, and key k. On defining $comp_X(v)$ and all_Y, respectively, we have

$$\Pr[Y = v] = \frac{|comp_Y(v)|}{all_Y}$$

all_Y is $2^n (2^n!)^{2^n} (2^n!)^{2^n}$, that is the number of keys times, the number of block ciphers. We next computed $|comp_X(v)|$ and $|comp_Y(v)|$. We knew that the view v contains the key k value, that is, at the end of the interaction, it is disclosed to distinguisher D. A set of input outputs of underlying block cipher E are derived and separately stored in tables T^1, T^2, and T^3. The number of input-output of E with the key value i is denoted as α_i and β_i in T^2 and T^3, respectively, where $0 \leq i \leq 2^n - 1$. The γ_i denotes the number of queries to O_1 with key value. There is no collision between any two tables, so v is good. Secondly, the distinguisher D never makes duplicate queries. Therefore, all the inputs and outputs of E in T^1, T^2, and T^3 are distinct, showing that $\gamma_i = \alpha_i$. The query response (u_1^i, w_1^i) of E in T^1 has $u_1^i = k$ or $u_1^i = 0$ (E1 to E20 have $u_1^i = k$ and others $u_1^i = 0$). On assuming $u_1^i = k$, we got

$$|comp_X(v)| = (2^n - \alpha_k - \beta_k - 1)! \prod_{i=0}^{k-1}(2^n - \alpha_i - \beta_i)! \prod_{i=k+1}^{2^n-1}(2^n - \alpha_i - \beta_i)!$$

$$|comp_Y(v)| = \prod_{i=0}^{2^n-1}(2^n - \gamma_i)! \left((2^n - \beta_k - 1)! \prod_{i=0}^{k-1}(2^n - \beta_i)! \prod_{i=k+1}^{2^n-1}(2^n - \beta_i)! \right)$$

$$= \prod_{i=0}^{2^n-1}(2^n - \alpha_i)! \left((2^n - \beta_k - 1)! \prod_{i=0}^{k-1}(2^n - \beta_i)! \prod_{i=k+1}^{2^n-1}(2^n - \beta_i)! \right)$$

$$= (2^n - \alpha_k)!(2^n - \beta_k - 1)! \prod_{i=0}^{k-1}(2^n - \alpha_i)!(2^n - \beta_i)! \prod_{i=k+1}^{2^n-1}(2^n - \alpha_i)!(2^n - \beta_i)!$$

From $(2^n - \alpha)!(2^n - \beta)! \leq (2^n - \alpha - \beta)!(2^n)!$, we have

$$|comp_Y(v)| \leq (2^n - \alpha_k - \beta_k - 1)! \, (2^n!)^{2^n}$$

We can compute

$$\frac{|comp_X(v)|}{|comp_Y(v)|} \geq \frac{(2^n - \alpha_k - \beta_k - 1)! \prod_{i=0}^{k-1}(2^n - \alpha_i - \beta_i)! \prod_{i=k+1}^{2^n-1}(2^n - \alpha_i - \beta_i)!}{(2^n - \alpha_k - \beta_k - 1)! \, (2^n!)^{2^n} \prod_{i=0}^{k-1}(2^n - \alpha_i - \beta_i)! \prod_{i=k+1}^{2^n-1}(2^n - \alpha_i - \beta_i)!} = \frac{1}{(2^n!)^{2^n}}$$

Finally, we can compute

$$\frac{\Pr[X = v]}{\Pr[X = v]} = \frac{|comp_X(v)|}{|comp_Y(v)|} \times \frac{all_Y}{all_X}$$

$$\geq \frac{1}{(2^n!)^{2^n}} \times \frac{2^n (2^n!)^{2^n} (2^n!)^{2^n}}{2^n (2^n!)^{2^n}} = 1$$

Thus, it gives a ratio for $V_{good} = 0$
Combining both 4.2 and 4.3,

$$Adv_{E1}^{prp}(q) \leq \frac{4q}{2^{n-1}}$$

Author Contributions: L.W. conceptualized the idea, Y.N. performed analysis, and both the authors wrote manuscript in coordination with each other.

Funding: National Nature Science Foundation of China, Youth Project.

Conflicts of Interest: The authors declare no conflict of interest.

References

1. Iwata, T. New Blockcipher Modes of Operation with Beyond the Birthday Bound Security. In *International Workshop on Fast Software Encryption*; Springer: Berlin/Heidelberg, Germany, 2006; pp. 310–327.
2. Patarin, J. Mirror theory and cryptography. *Appl. Algebra Eng. Commun. Comput.* **2017**, *28*, 321–338. [CrossRef]
3. Iwata, T.; Mennink, B.; Vizár, D. Cenc is optimally secure. *IACR Cryptol. ePrint Arch.* **2016**, *2016*, 1087.
4. Bellare, M.; Desai, A.; Jokipii, E.; Rogaway, P. A concrete security treatment of symmetric encryption. In Proceedings of the 38th Annual Symposium on Foundations of Computer Science, Miami Beach, FL, USA, 20–22 October 1997; pp. 394–403.
5. Bellare, M.; Guérin, R.; Rogaway, P. Xor macs: New methods for message authentication using finite pseudorandom functions. In *Annual International Cryptology Conference*; Springer: Berlin/Heidelberg, Germany, 1995; pp. 15–28.
6. Bernstein, D.J. How to stretch random functions: The security of protected counter sums. *J. Cryptol.* **1999**, *12*, 185–192. [CrossRef]

7. McGrew, D.A.; Viega, J. The security and performance of the galois/counter mode (gcm) of operation. In *International Conference on Cryptology in India*; Springer: Berlin/Heidelberg, Germany, 2004; pp. 343–355.
8. Patarin, J. A Proof of Security in $O_{(2^n)}$ for the Xor of Two Random Permutations. In *International Conference on Information Theoretic Security*; Springer: Berlin/Heidelberg, Germany, 2008; pp. 232–248.
9. Patarin, J. Luby-rackoff: 7 rounds are enough for $2^{n(1-\varepsilon)}$ security. In *Annual International Cryptology Conference*; Springer: Berlin/Heidelberg, Germany, 2003; pp. 513–529.
10. Patarin, J. On linear systems of equations with distinct variables and small block size. In *International Conference on Information Security and Cryptology*; Springer: Berlin/Heidelberg, Germany, 2005; pp. 299–321.
11. Patarin, J. Introduction to mirror theory: Analysis of systems of linear equalities and linear non equalities for cryptography. *IACR Cryptol. ePrint Arch.* **2010**, *2010*, 287.
12. Daemen, J.; Rijmen, V. Rijndael/aes. *Encycl. Cryptogr. Secur.* **2005**, 520–524. [CrossRef]
13. Bogdanov, A.; Knudsen, L.R.; Leander, G.; Paar, C.; Poschmann, A.; Robshaw, M.J.; Seurin, Y.; Vikkelsoe, C. Present: An ultra-lightweight block cipher. In *International Workshop on Cryptographic Hardware and Embedded Systems*; Springer: Berlin/Heidelberg, Germany, 2007; pp. 450–466.
14. De Canniere, C.; Dunkelman, O.; Knežević, M. Katan and ktantan—A family of small and efficient hardware-oriented block ciphers. In *International Workshop on Cryptographic Hardware and Embedded Systems*; Springer: Berlin/Heidelberg, Germany, 2009; pp. 272–288.
15. Guo, J.; Peyrin, T.; Poschmann, A.; Robshaw, M. The led block cipher. In *International Workshop on Cryptographic Hardware and Embedded Systems*; Springer: Berlin/Heidelberg, Germany, 2011; pp. 326–341.
16. Impagliazzo, R.; Rudich, S. Limits on the provable consequences of one-way permutations (invited talk). In *Proceedings on Advances in Cryptology*; Springer: Berlin/Heidelberg, Germany, 1990; pp. 8–26.
17. Hall, C.; Wagner, D.; Kelsey, J.; Schneier, B. Building prfs from prps. In *Annual International Cryptology Conference*; Springer: Berlin/Heidelberg, Germany, 1998; pp. 370–389.
18. Bellare, M.; Rogaway, P. The security of triple encryption and a framework for code-based game-playing proofs. In *Annual International Conference on the Theory and Applications of Cryptographic Techniques*; Springer: Berlin/Heidelberg, Germany, 2006; pp. 409–426.
19. Chang, D.; Nandi, M. A short proof of the prp/prf switching lemma. *IACR Cryptol. ePrint Arch.* **2008**, *2008*, 78.
20. Bellare, M.; Krovetz, T.; Rogaway, P. Luby-rackoff backwards: Increasing security by making block ciphers non-invertible. In *International Conference on the Theory and Applications of Cryptographic Techniques*; Springer: Berlin/Heidelberg, Germany, 1998; pp. 266–280.
21. Lucks, S. The sum of prps is a secure prf. In *International Conference on the Theory and Applications of Cryptographic Techniques*; Springer: Berlin/Heidelberg, Germany, 2000; pp. 470–484.
22. Lim, C.H.; Korkishko, T. Mcrypton–a lightweight block cipher for security of low-cost rfid tags and sensors. In *International Workshop on Information Security Applications*; Springer: Berlin/Heidelberg, Germany, 2005; pp. 243–258.
23. Wu, W.; Zhang, L. *Lblock: A Lightweight Block Cipher*; Springer: Berlin/Heidelberg, Germany, 2011; pp. 327–344.
24. Borghoff, J.; Canteaut, A.; Güneysu, T.; Kavun, E.B.; Knezevic, M.; Knudsen, L.R.; Leander, G.; Nikov, V.; Paar, C.; Rechberger, C.; et al. *Prince—A Low-Latency Block Cipher for Pervasive Computing App. Lications*; Springer: Berlin/Heidelberg, Germany, 2012; pp. 208–225.
25. Beaulieu, R.; Treatman-Clark, S.; Shors, D.; Weeks, B.; Smith, J.; Wingers, L. The simon and speck lightweight block ciphers. In Proceedings of the 52nd ACM/EDAC/IEEE Design Automation Conference (DAC), San Francisco, CA, USA, 8–12 June 2015; pp. 1–6.
26. Beierle, C.; Jean, J.; Kölbl, S.; Leander, G.; Moradi, A.; Peyrin, T.; Sasaki, Y.; Sasdrich, P.; Sim, S.M. The skinny family of block ciphers and its low-latency variant mantis. In *Annual International Cryptology Conference*; Springer: Berlin/Heidelberg, Germany, 2016; pp. 123–153.
27. Bellare, M.; Impagliazzo, R. A tool for obtaining tighter security analyses of pseudorandom function based constructions, with app. lications to prp to prf conversion. *IACR Cryptol. ePrint Arch.* **1999**, *1999*, 24.
28. Patarin, J. Security in $O(2^n)$ for the xor of two random permutations\-proof with the standard h technique. *IACR Cryptol. ePrint Arch.* **2013**, *2013*, 368.
29. Cogliati, B.; Lampe, R.; Patarin, J. The indistinguishability of the xor of $$ k $$ permutations. In *International Workshop on Fast Software Encryption*; Springer: Berlin/Heidelberg, Germany, 2014; pp. 285–302.
30. Dai, W.; Hoang, V.T.; Tessaro, S. Information-theoretic indistinguishability via the chi-squared method. In *Annual International Cryptology Conference*; Springer: Berlin/Heidelberg, Germany, 2017; pp. 497–523.

31. Bhattacharya, S.; Nandi, M. Revisiting variable output length xor pseudorandom function. *IACR Trans. Symmetric Cryptol.* **2018**, *2018*, 314–335.
32. Yasuda, K. A new variant of pmac: Beyond the birthday bound. In *Annual Cryptology Conference*; Springer: Berlin/Heidelberg, Germany, 2011; pp. 596–609.
33. Datta, N.; Dutta, A.; Nandi, M.; Paul, G.; Zhang, L. Single key variant of PMAC_plus. *IACR Trans. Symmetric Cryptol.* **2017**, *2017*, 268–305.
34. Naito, Y. Blockcipher-based macs: Beyond the birthday bound without message length. In *International Conference on the Theory and App.lication of Cryptology and Information Security*; Springer: Berlin/Heidelberg, Germany, 2017; pp. 446–470.
35. Gilboa, S.; Gueron, S. The advantage of truncated permutations. In *International Symposium on Cyber Security Cryptography and Machine Learning*; Springer: Berlin/Heidelberg, Germany, 2019; pp. 111–120.
36. Cogliati, B.; Seurin, Y. Ewcdm: An efficient, beyond-birthday secure, nonce-misuse resistant mac. In *Annual International Cryptology Conference*; Springer: Berlin/Heidelberg, Germany, 2016; pp. 121–149.
37. Mennink, B.; Neves, S. Encrypted davies-meyer and its dual: Towards optimal security using mirror theory. In *Annual International Cryptology Conference*; Springer: Berlin/Heidelberg, Germany, 2017; pp. 556–583.
38. Mennink, B.; Neves, S. Optimal prfs from blockcipher designs. *IACR Trans. Symmetric Cryptol.* **2017**, 228–252.
39. Chen, Y.L.; Lambooij, E.; Mennink, B. How to build pseudorandom functions from public random permutations. In *Annual International Cryptology Conference*; Springer: Berlin/Heidelberg, Germany, 2019; pp. 266–293.
40. Beaulieu, R.; Shors, D.; Smith, J.; Treatman-Clark, S.; Weeks, B.; Wingers, L. Simon and speck: Block ciphers for the internet of things. *IACR Cryptol. ePrint Arch.* **2015**, *2015*, 585.
41. Chen, S.; Steinberger, J. Tight security bounds for key-alternating ciphers. In *Annual International Conference on the Theory and Applications of Cryptographic Techniques*; Springer: Berlin/Heidelberg, Germany, 2014; pp. 327–350.
42. Even, S.; Mansour, Y. A construction of a cipher from a single pseudorandom permutation. *J. Cryptol.* **1997**, *10*, 151–161. [CrossRef]

© 2019 by the authors. Licensee MDPI, Basel, Switzerland. This article is an open access article distributed under the terms and conditions of the Creative Commons Attribution (CC BY) license (http://creativecommons.org/licenses/by/4.0/).

Article

Ideals of Numerical Semigroups and Error-Correcting Codes

Maria Bras-Amorós

Department of Computer Science and Mathematics, Universitat Rovira i Virgili, 43007 Tarragona, Catalonia, Spain; maria.bras@urv.cat

Received: 8 October 2019; Accepted: 11 November 2019; Published: 14 November 2019

Abstract: Several results relating additive ideals of numerical semigroups and algebraic-geometry codes are presented. In particular, we deal with the set of non-redundant parity-checks, the code length, the generalized Hamming weights, and the isometry-dual sequences of algebraic-geometry codes from the perspective of the related Weierstrass semigroups. These results are related to cryptographic problems such as the wire-tap channel, t-resilient functions, list-decoding, network coding, and ramp secret sharing schemes.

Keywords: numerical semigroup; Weierstrass semigroup; semigroup ideal; error-correcting code; algebraic-geometry code

1. Introduction

In a previous survey chapter [1], numerical semigroups were presented together with some of the related classical problems, and their importance for algebraic-geometry codes was explained. In particular, numerical semigroups can be used to establish decoding conditions, are useful to define bounds for the minimum distance of codes, and to improve the code dimension. In this contribution, which is a continuation of that chapter, we will present some results relating ideals of numerical semigroups and the set of non-redundant parity-checks, the code length, the generalized Hamming weights, and the isometry-dual sequences of algebraic-geometry codes. The reader not familiar with algebraic geometry may be interested in the introductory sections of [1].

The organization of this contribution is as follows. Section 2 introduces numerical semigroups and states basic notions, in particular the Frobenius number and symmetric semigroups, which will be important in the following sections. Section 3 presents ideals of numerical semigroups and some results connecting the maximum gap of an ideal with the size of the complement of the ideal. Maximum sparse ideals are defined as those ideals for which this maximum gap is maximum restricted to a given size of the complement, and this connects with symmetric semigroups. Section 4 presents one-point algebraic-geometry codes and relates redundant checks with ideals of numerical semigroups. Section 5 deals with the Geil–Matsumoto bound for the number of points a curve can have and so with the length of codes. Section 6 deals with the sequences of one-point algebraic-geometry codes that satisfy the isometry-dual property and the effects of puncturing such sequences. The results are derived from the results on maximum sparse ideals of numerical semigroups. Section 7 deals with the generalized Hamming weights of algebraic-geometry codes by means of Feng–Rao numbers and Weierstrass semigroups. These results are related to cryptographic problems such as the wire-tap channel or ramp secret sharing schemes.

2. Numerical Semigroups

2.1. Basic Notions

A *numerical semigroup* is a subset Λ of $\mathbb{N}_0$ that contains 0, contains any finite sum of its elements, and its complement in $\mathbb{N}_0$ is finite. Weierstrass semigroups are indeed numerical semigroups.

The *genus* of a numerical semigroup Λ is the amount $g = \#(\mathbb{N}_0 \setminus \Lambda)$. The elements belonging to the semigroup Λ are its *non-gaps* while the positive elements in its complement are its *gaps*. There is a unique increasing bijective map $\lambda : \mathbb{N}_0 \longrightarrow \Lambda$. We call it the *enumeration* of Λ, and the notation λ_i will be used for $\lambda(i)$.

The *generators* of a semigroup are those nonzero elements in the semigroup that are not the result of adding two other nonzero elements in the semigroup. The whole set of generators is necessarily finite and coprime. Conversely, if a finite set G of positive integers is coprime, the set of finite sums of the elements in G is called the *semigroup generated* by G and it is denoted by $\langle G \rangle$.

2.2. Frobenius Number and Symmetric Semigroups

The *conductor* of a numerical semigroup Λ is the least integer in the semigroup for which all integers larger than it belong to the semigroup. The conductor minus one is then the maximum gap of the numerical semigroup, which is called the *Frobenius number* of the semigroup. It can be easily proved using the Pigeonhole Principle that the conductor is at most twice the genus. The semigroups that attain this bound are called *symmetric semigroups*. The symmetry of a semigroup Λ comes from the fact that, if the Frobenius number F and the genus g of the semigroup satisfy $F = 2g - 1$, then the semigroup satisfies $i \in \Lambda \iff F - i \notin \Lambda$.

2.3. Semigroups Generated by Two Integers

Weierstrass semigroups generated by two integers are very common as is the case in hyperelliptic curves or Geil's norm-trace curves [2]. Most important, for any coprime positive integers a and b, one can find a curve with a point whose Weierstrass semigroup is $\langle a, b \rangle$ [3].

Sylvester's formula [4] states that the Frobenius number of the semigroup $\langle a, b \rangle$ is $ab - a - b$, while its genus is $\frac{(a-1)(b-1)}{2}$. Hence, semigroups generated by two positive integers satisfy the symmetry property.

Example 1 (Hermitian curve $\mathcal{H}_q$). Let q be a prime power. The Hermitian curve $\mathcal{H}_q$ over $\mathbb{F}_{q^2}$ is defined by the affine equation $x^{q+1} = y^q + y$ and homogeneous equation $X^{q+1} - Y^q Z - YZ^q = 0$. The point $P_\infty = (0 : 1 : 0)$ is the unique point of $\mathcal{H}_q$ at infinity. It can be proved (see, for instance, [1]) that $v_{P_\infty}(\frac{Z}{Y}) = q + 1$ and $v_{P_\infty}(\frac{X}{Z}) = -q$. Hence, the Weierstrass semigroup Λ at P_∞ contains the semigroup generated by $q, q+1$ whose complement in $\mathbb{N}_0$ has $\frac{q(q-1)}{2} = g$ elements. Since we know that the complement of Λ in $\mathbb{N}_0$ also has g elements, this means that both semigroups are the same. For further details on the Hermitian curve, see [3,5].

3. Ideals of Numerical Semigroups

3.1. Ideals

A subset I of a numerical semigroup Λ is an *ideal* of Λ if $I + \Lambda \subseteq I$. We say that I is a *proper* ideal of Λ if $I \neq \Lambda$. Because of the finiteness of the complement of Λ and the definition of an ideal, the complement of an ideal (either with respect to the ideal or with respect to $\mathbb{N}_0$) must be finite as well. Hence, we can consider the largest integer in the complement of an ideal (with respect to $\mathbb{N}_0$). It is called the *Frobenius number* of the ideal.

Next, we will prove an upper bound on the Frobenius number of an ideal which extends the upper bound for the Frobenius number of a numerical semigroup that is twice the genus minus one. Indeed, we will see that the Frobenius number of an ideal is less than or equal to the number of elements in the semigroup which do not belong to the ideal plus the double of the genus of the semigroup minus one. Notice that, if we take the ideal to be the whole semigroup, then we get the already known bound for the Frobenius number of a numerical semigroup. Hence, this result, stated in Theorem 1, can be seen as a generalization of the upper bound for the Frobenius number of the numerical semigroup. The ideals for which the Frobenius number attains the bound are called *maximum sparse* ideals. All the results in this section were first proved in [6].

3.2. The Frobenius Number of an Ideal

Suppose that Λ is a numerical semigroup and that I is an ideal of Λ. The *difference* of the ideal I with respect to Λ is the number of elements in $\Lambda \setminus I$. One can prove (see [3], Lemma 5.15) that, in the case of principal ideals, that is, ideals of the form $a + \Lambda$ for some nonnegative integer a, the difference is exactly a. From this, it is straightforward to deduce that the Frobenius number of $a + \Lambda$ is less than or equal to twice the genus of the semigroup plus a (which is the difference) minus one. This will be generalized to the bound in Theorem 1 for any ideal. Furthermore, the semigroups for which the bound is attained will be characterized.

3.3. Upper Bounding the Frobenius Number of an Ideal

For each nonnegative integer i, define $D(i) = \{\lambda_j \leq \lambda_i : \lambda_i - \lambda_j \in \Lambda\}$. The set $D(i)$ is often called the set of divisors of λ_i, and its cardinality is denoted $\nu_i = \#D(i)$. The sequence ν_i has many implications in coding theory. It is fundamental in the computation of bounds for the minimum distance of algebraic-geometry codes based on a single point as well as in the optimization of the redundancy of those codes. Its properties and applications can be seen in [7–14] and in the survey [1]. As a curiosity, it was proved in [7,15] that the set of elements of a numerical semigroup is determined by its ν sequence. However, it was proved in [8] that, given a finite subset of values of the ν sequence, it is contained in the ν sequence of infinitely many numerical semigroups. We will see how the sets $D(i)$ are related to ideals of semigroups. Next, we present two lemmas proved by Barucci in [16] and by Høholdt, van Lint, and Pellikaan in [3], respectively, and the main theorem that can be derived from the two lemmas.

Lemma 1 ([16]). *Every ideal of a numerical semigroup Λ can be expressed as an intersection of finitely many irreducible ideals and irreducible ideals are expressible as $\Lambda \setminus D(i)$ for some i.*

Lemma 2 (([3] Theorem 5.24)). *Let $g(i)$ be the number of gaps in the interval from 1 to $\lambda_i - 1$ and let $G(i)$ be the number of pairs of gaps whose sum equals λ_i. Then, $\nu_i = i - g(i) + G(i) + 1$.*

Theorem 1. *Suppose that I is an ideal of a numerical semigroup of genus g so that $\Lambda \setminus I$ has d elements. Then, $d + 2g + i \in I$ for all $i \geq 0$. Equivalently, the Frobenius number of I is less than or equal to $d + 2g - 1$.*

Proof. It is straightforward to see that the intersection of two ideals satisfying the result also satisfies the result. Now, by Lemma 1, it will be enough to show that the result holds for the ideals expressible as $I = \Lambda \setminus D(i)$. Equivalently, $\nu_i + 2g \geq \max\{c, \lambda_i + 1\}$, with c the conductor of Λ. This holds if $c \geq \lambda_i + 1$ since $c \leq 2g$. Otherwise, if $\lambda_i + 1 > c$, then $g(i) = g$, $\lambda_i = i + g$, and as a consequence of Lemma 2, $\nu_i + 2g = (i - g + G(i) + 1) + 2g = i + g + 1 + G(i) = \lambda_i + 1 + G(i) > \lambda_i + 1$. □

The ideals for which the Frobenius number attains the previous bound will be called *maximum sparse* ideals.

3.4. Maximum Sparse Ideals

In next theorem, we characterize the ideals that are maximum sparse.

Theorem 2. *The statements that follow are equivalent for an ideal I with difference $d > 0$ of a semigroup Λ with genus g:*

1. *The Frobenius number of the ideal I equals $d + 2g - 1$.*
2. *$I = \Lambda \setminus D(i)$ for some i such that $G(i) = 0$.*

Proof. On one hand, let the Frobenius number of the ideal I be $d + 2g - 1$. If I is a non-trivial intersection of the ideals I' and I'', whose differences are, respectively, d' and d'', then the difference d of I is strictly larger than both d' and d''. If $d + 2g - 1$ is not an element of I, then it is neither an element of I' nor an element of I'', but the value $d + 2g - 1$ is strictly larger than both $d' + 2g - 1$ and $d'' + 2g - 1$. This contradicts Theorem 1. This implies, by Lemma 1, that I is of the form $\Lambda \setminus D(i)$ for some i. Now, $d = v_i$ because $I = \Lambda \setminus D(i)$. If λ_i is smaller than c, then $v_i + 2g - 1 \geq = 2g \geq c$, hence $d + 2g - 1 \in I$, contradicting our assumption. Consequently, $\lambda_i \geq c$ and by Lemma 2, $v_i = i - g + G(i) + 1$. Thus, $d + 2g - 1 = i + g + G(i) = \lambda_i + G(i)$. However, $d + 2g - 1 \notin I$, and so $G(i) = 0$.

On the other hand, suppose I is of the form $\Lambda \setminus D(i)$ for some i with $G(i) = 0$, and so $d = v_i$. By the former remarks, since $G(i) = 0$, one deduces that $\lambda_i = i + g$ and, by Lemma 2, it follows that $d + 2g - 1 = \lambda_i \notin I$. □

Example 2 (Weierstrass semigroup of $\mathcal{H}_4$). *The Weierstrass semigroup of $\mathcal{H}_4$ is $\Lambda = \{0, 4, 5, 8, 9, 10, 12, 13, \dots\}$. We wish to find all the maximum sparse ideals of Λ. Since the Frobenius number of Λ is 11 and $11 + 11 = 22 = \lambda_{16}$, it holds that $G(16) > 0$ while $G(i) = 0$ for all $i \geq 17$. This implies that all ideals of the form $\Lambda \setminus D(i)$ with $i \geq 17$ are maximum sparse. Let us see now whether $G(i) = 0$ for all i with $6 \leq i \leq 15$. On one hand, $G(6) > 0$ since $\lambda_6 = 12 = 11 + 1$; $G(7) > 0$ since $\lambda_7 = 13 = 11 + 2$; $G(8) > 0$ since $\lambda_8 = 14 = 11 + 3$; $G(9) = 0$ because the difference between 15 and any gap is a non-gap, indeed, $\{15 - 1 = 14, 15 - 2 = 13, 15 - 3 = 12, 15 - 6 = 9, 15 - 7 = 8, 15 - 11 = 4\} \subseteq \Lambda$; $G(10) = 0$ because the difference between 16 and any gap is a non-gap, indeed, $\{16 - 1 = 15, 16 - 2 = 14, 16 - 3 = 13, 16 - 6 = 10, 16 - 7 = 9, 16 - 11 = 5\} \subseteq \Lambda$; $G(11) > 0$ since $\lambda_{11} = 17 = 11 + 6$; $G(12) > 0$ since $\lambda_i = 18 = 11 + 7$; $G(13) = 0$ because the difference between 19 and any gap is a non-gap, indeed, $\{19 - 1 = 18, 19 - 2 = 17, 19 - 3 = 16, 19 - 6 = 13, 19 - 7 = 12, 19 - 11 = 8\} \subseteq \Lambda$. $G(14) = 0$ because the difference between 20 and any gap is a non-gap, indeed, $\{20 - 1 = 19, 20 - 2 = 18, 20 - 3 = 17, 20 - 6 = 14, 20 - 7 = 13, 20 - 11 = 9\} \subseteq \Lambda$. $G(15) = 0$ because the difference between 21 and any gap is a non-gap, indeed, $\{21 - 1 = 20, 21 - 2 = 19, 21 - 3 = 18, 21 - 6 = 15, 21 - 7 = 14, 21 - 11 = 10\} \subseteq \Lambda$.*

Hence, all maximum sparse ideals are $I_9 = \Lambda \setminus D(9) = \{4, 8, 9, 12, 13, 14, 16, 17, 18, 19, 20, 21, 22, \dots\}$, where $D(9) = \{0, 5, 10, 15\}$, $d = 4$, and $d + 2g - 1 = 15$; $I_{10} = \Lambda \setminus D(10) = \{5, 9, 10, 13, 14, 15, 17, 18, 19, 20, 21, 22, \dots\}$, where $D(10) = \{0, 4, 8, 12, 16\}$, $d = 5$, and $d + 2g - 1 = 16$; $I_{13} = \Lambda \setminus D(13) = \{8, 12, 13, 16, 17, 18, 20, 21, 22, \dots\}$, where $D(13) = \{0, 4, 5, 9, 10, 14, 15, 19\}$, $d = 8$, and $d + 2g - 1 = 19$; $I_{14} = \Lambda \setminus D(14) = \{9, 13, 14, 17, 18, 19, 21, 22, \dots\}$, where $D(14) = \{0, 4, 5, 8, 10, 12, 15, 16, 20\}$, $d = 9$, and $d + 2g - 1 = 20$; $I_{15} = \Lambda \setminus D(15) = \{10, 14, 15, 18, 19, 20, 22, \dots\}$, where $D(15) = \{0, 4, 5, 8, 9, 12, 13, 16, 17, 21\}$, $d = 10$, and $d + 2g - 1 = 21$; $I_{17} = \Lambda \setminus D(17) = \{12, 16, 17, 20, 21, 22, 24, \dots\}$, where $D(17) = \{0, 4, 5, 8, 9, 10, 13, 14, 15, 18, 19, 23\}$, $d = 12$, and $d + 2g - 1 = 23$; and finally $\Lambda \setminus D(i)$ for all $i > 17$. Here, $D(i) = \{0, 4, 5, 8, 9, 10, 12, 13, \dots, i + 6 - 12, i + 6 - 10, i + 6 - 9, i + 6 - 8, i + 6 - 5, i + 6 - 4, i + 6\}$, $d = i - 5$, and $d + 2g - 1 = i + 6$.

The next corollary characterizes maximum sparse ideals of symmetric semigroups.

Corollary 1. *Maximum sparse ideals of a symmetric semigroup are exactly the principal ideals of the semigroup.*

Proof. It has already been explained that the difference of the principal ideal $a + \Lambda$ is exactly a, and so it is obvious that principal ideals of symmetric semigroups are maximum sparse.

Suppose now that I is a maximum sparse ideal of a symmetric semigroup Λ. If $I = \Lambda$, the result is obvious. Otherwise, by Theorem 2, $I = \Lambda \setminus D(i)$ for some i with $G(i) = 0$. Let a be the minimum element of I. Since $I = \Lambda \setminus D(i)$, the difference $\lambda_i - a$ is a gap of Λ. By the minimality of a, the gap $\lambda_i - a$ must be the Frobenius number F of Λ since, otherwise, $\lambda_i - F$ would be an element in Λ (because $G(i) = 0$) not in $D(i)$ and, so, an element of I smaller than a. Now, it remains to see that any element $\mu \in \Lambda \setminus D(i)$ belongs to $a + \Lambda$. Indeed, $\mu - a = \mu - (\lambda_i - F) = F - (\lambda_i - \mu)$. Since $\mu \notin D(i)$, we have $(\lambda_i - \mu) \notin \Lambda$ and, by the symmetry of Λ, we have $F - (\lambda_i - \mu) \in \Lambda$. Thus, $\mu - a \in \Lambda$. □

Example 3 (Weierstrass semigroup of $\mathcal{H}_4$). *The Weierstrass semigroup of the point at infinity of $\mathcal{H}_4$ is generated by two integers and so it is symmetric. The previous corollary in this case can be checked for the set of maximum sparse ideals listed in Example 2.*

Remark 1. *It is important to remark that the hypothesis in Corollary 1 is necessary. A counterexample can be found in the semigroup $\Lambda = \{0, 4, 8, 9, \dots\}$, of Frobenius number 7 and genus 6, and so, not symmetric. The semigroup Λ has the ideal $I = \{9, 10, 11, 13, 14, 15, 17, \dots\}$, which equals $\Lambda \setminus D(10) = \Lambda \setminus \{0, 4, 8, 12, 16\}$. The ideal I has difference $d = 5$ and Frobenius number $16 = d + 2g - 1$. Hence, I is a maximum sparse ideal, but it is not principal because it is, indeed, $I = (9 + \Lambda) \cup \{10, 11, 14, 15\}$.*

3.5. The Ideal of Frobenius Numbers of Sparse Ideals

The next lemma shows that the Frobenius numbers of the maximum sparse ideals of a numerical semigroup constitute in turn another ideal of the numerical semigroup.

Lemma 3. *The nonzero non-gaps λ_i such that $G(i) = 0$ constitute an ideal L of Λ.*

Proof. First of all, notice that $G(i) = 0$ is not satisfied if λ_i is smaller than the conductor. Indeed, if λ_i is smaller than the conductor c, then there must be a gap a smaller than λ_i with $\lambda_i - a < \lambda_1$, since, otherwise, λ_i would not be smaller than the conductor. Now, $\lambda_i - a$ must be a positive gap and $\lambda_i = (\lambda_i - a) + a$, a contradiction with $G(i) = 0$. Hence, the elements in L are equal than or equal to the conductor of Λ.

It remains to show that, if $\lambda_i \in L$, then $\lambda_i + \lambda_j \in L$ for any $\lambda_j \in \Lambda$. Assume that $\lambda_j \neq 0$. Let k be such that $\lambda_i + \lambda_j = \lambda_k$. Suppose that $\lambda_k \notin L$, that is, $G(k) \neq 0$. Then, there are two gaps a, a' with $\lambda_k = a + a'$. Note that both $a, a' < \lambda_i = \lambda_k - \lambda_j$ since λ_i is greater than or equal to c. From $a + a' = \lambda_k$, we have $\lambda_j < a, a' < \lambda_i$. Then, $a - \lambda_j$ does not belong to Λ because, otherwise, $a = \lambda_j + (a - \lambda_j) \in \Lambda + \Lambda \subseteq \Lambda$. In particular, $(a - \lambda_j) + a'$ is a sum of two gaps equal to $a + a' - \lambda_j = \lambda_k - \lambda_j = \lambda_i$, a contradiction with $G(i) = 0$. □

4. One-Point Algebraic-Geometry Codes

In coding theory, by a *linear code* of length n, it is meant a linear subspace C of $\mathbb{F}_q^n$, with $\mathbb{F}_q$ the field of order q, for some prime power q. Its dimension is usually denoted k. The *dual code* of a linear code is its orthogonal space. It has the same length than C and dimension $n - k$. A knowledge of the dual code is useful in most decoding algorithms. To compare two different vectors of $\mathbb{F}_q^n$, one counts the number of differing positions and this number is referred to as the *Hamming distance* between the two vectors. The *weight* of a vector is defined as its Hamming distance to the all-zero vector. An important parameter of a code is its *minimum distance*, representing the minimum of the Hamming distances between each pair of different vectors in the code. The *correction capability* of a code tells how far we can go from any code vector with the guarantee that we will not get closer to a code vector different than the originary one. The correction capability is exactly $\lfloor \frac{d-1}{2} \rfloor$ if the minimum distance of the code is d.

An important class of error-correcting codes are the algebraic-geometry codes. Let $\mathcal{X}$ be a smooth irreducible algebraic curve over $\mathbb{F}_q$ and let Q be a rational point of $\mathcal{X}$. Let Λ be the Weierstrass semigroup at Q and let $A = \bigcup_{m \geq 0} L(mQ)$ be the ring of rational functions of $\mathcal{X}$ only having poles at Q. There exists a basis $z_0, z_1, \dots, z_i, \dots$ of A such that $v_Q(z_i) = -\lambda_i$. Now, for each collection of rational points $P_1, \ldots, P_n$, all of them different from Q, and each set of indices $B \subseteq \mathbb{N}_0$, define the *one-point code* $C_B = <(z_i(P_1), \ldots, z_i(P_n)) : i \in B>$. The elements in the set B are called *parity checks* of C_B and the one-point code is said to be classical if $B = \{0, 1, \ldots, m\}$. We will use C_m to refer to $C_{\{0,\ldots,m\}}$. In the present survey, we consider only the codes C_m. Ref. [1] is a survey on results related to the minimum distance, the error-correction capability, and the redundancy of the codes C_B from the perspective of Weierstrass semigroups. In that case we considered, though, the dual codes $<(z_i(P_1), \ldots, z_i(P_n)) : i \in B>^\perp$.

It can be shown that $C_m = \{(f(P_1), \ldots, f(P_n)) : f \in L(\lambda_m Q)\}$. Note that it can be the case that $C_m = C_{m-1}$. The next lemma is stated in other words in ([17], Corollary 3.3).

Lemma 4. *Suppose that Λ is the Weierstrass semigroup at a rational point Q and define $\Lambda^* = \{0\} \cup \{m \in \mathbb{N}, m > 0 : C_m \neq C_{m-1}\} = \{m_0 = 0, m_1, \ldots, m_n\}$. Then, the set $\Lambda \setminus \Lambda^*$ is an ideal of Λ.*

5. Ideals and the Length of Algebraic-Geometry Codes

From the previous definition of algebraic-geometry codes, we see that the length of a code defined over an algebraic smooth irreducible curve is conditioned by the number of points of the curve. Thus, bounding the number of points of smooth irreducible curves becomes an important problem of algebraic-geometry codes.

5.1. The Geil–Matsumoto Bound

Define $N_q(g)$ as the maximum number of points an irreducible smooth curve of genus g can have over the finite field of q elements. The Hasse–Weil bound is $|N_q(g) - q - 1| \leqslant 2g\sqrt{q}$ ([18], Theorem V.2.3), which is refined by Serre's bound $|N_q(g) - q - 1| \leqslant g\lfloor 2\sqrt{q} \rfloor$ ([18], Theorem V.3.1). The web page [19] is devoted to give the best known examples of curves with many points for any fixed pair q, g.

Suppose that, for an irreducible smooth curve $\mathcal{X}$ over $\mathbb{F}_q$, we not only know its genus but also the Weierstrass semigroup Λ at a given point. We may wonder, with this assumption, how many points $\mathcal{X}$ can have. For this goal, we define $N_q(\Lambda)$ to be the maximum number of possible points. The first bound is due to Lewittes [20], and it uses only the first element λ_1 of Λ different than 0. It is $N_q(\Lambda) \leqslant L_q(\Lambda) := q\lambda_1 + 1$. On the other hand, Geil and Matsumoto [21] proved that

$$N_q(\Lambda) \leqslant GM_q(\Lambda) := \#(\Lambda \setminus \cup_{\lambda_i \text{ generator of } \Lambda}(q\lambda_i + \Lambda)) + 1. \quad (1)$$

Using the fact that

$$\#(\Lambda \setminus (q\lambda_1 + \Lambda)) = q\lambda_1, \quad (2)$$

proved in [3,21], one can deduce Lewittes' bound from the Geil–Matsumoto bound.

Remark 2. *The set $\Lambda \setminus \cup_{\lambda_i \text{ generator of } \Lambda}(q\lambda_i + \Lambda)$ is the complement of an ideal of Λ. Hence, any advance in the comprehension of ideals of numerical semigroups may result in new bounds for the length of algebraic-geometry codes.*

For a numerical semigroup generated by two coprime integers a, b, it can be proved [22] that the Geil–Matsumoto bound is exactly as follows:

$$GM_q(\langle a, b \rangle) = \sum_{n=0}^{a-1} \min\left(q, \left\lceil \frac{q-n}{a} \right\rceil \cdot b\right) + 1 \quad (3)$$

$$= \begin{cases} qa + 1 & \text{if } q \leqslant \lfloor \frac{q}{a} \rfloor b, \\ (q \bmod a)q + (a - (q \bmod a))\lfloor \frac{q}{a} \rfloor b + 1 & \text{if } \lfloor \frac{q}{a} \rfloor b < q \leqslant \lceil \frac{q}{a} \rceil b, \\ ab\lceil \frac{q}{a} \rceil - (a - (q \bmod a))b + 1 & \text{if } q > \lceil \frac{q}{a} \rceil b. \end{cases} \quad (4)$$

5.2. Coincidences of Lewittes's and the Geil–Matsumoto Bound

It was proved by Beelen and Ruano in ([23], Proposition 9) that, if $q \in \Lambda$, then the Lewittes and the Geil–Matsumoto bounds coincide. For two-generated semigroups, Equation (3) implies that both bounds coincide if and only if $q \leqslant \lfloor \frac{q}{a} \rfloor b$. Otherwise, the Lewittes bound is improved by the Geil–Matsumoto bound. This result for two-generated semigroups can be generalized to semigroups of any number of generators (larger than or equal to two). This is the goal of this subsection. The results are taken from [22].

Theorem 3. *Let $\Lambda = \langle \lambda_1, \ldots, \lambda_n \rangle$ with $\lambda_1 < \lambda_i$ for all $i > 1$. The next statements are equivalent*

1. $GM_q(\Lambda) = L_q(\Lambda)$;
2. $\Lambda \setminus \cup_{i=1}^{n}(q\lambda_i + \Lambda) = \Lambda \setminus (q\lambda_1 + \Lambda)$;
3. $q(\lambda_i - \lambda_1) \in \Lambda$ for all $i > 1$.

Proof. By Equation (2), it is straightforward to prove that 2 implies 1. The reverse implication follows from the inclusion $\Lambda \setminus \cup_{i=1}^{n}(q\lambda_i + \Lambda) \subseteq \Lambda \setminus (q\lambda_1 + \Lambda)$ and the equality $GM_q(\Lambda) = L_q(\Lambda)$, which, by Equation (2), implies that $\#(\Lambda \setminus \cup_{i=1}^{n}(q\lambda_i + \Lambda)) = \#(\Lambda \setminus (q\lambda_1 + \Lambda))$.

For the equivalence of the last two statements, notice that $q(\lambda_i - \lambda_1) \in \Lambda$ for all $i > 1 \iff q\lambda_i \in q\lambda_1 + \Lambda$ for all $i > 1 \iff q\lambda_i + \Lambda \subseteq q\lambda_1 + \Lambda$ for all $i > 1 \iff \Lambda \setminus \cup_{i=1}^{n}(q\lambda_i + \Lambda) = \Lambda \setminus (q\lambda_1 + \Lambda)$. □

Notice that Theorem 3 implies Beelen–Ruano's result since $q \in \Lambda$ implies $q(\lambda - \lambda_1) \in \Lambda$ for all $\lambda \in \Lambda$.

From Theorem 3, it makes sense to analyze the conditions under which $q(\lambda_i - \lambda_1) \in \Lambda$ for some $i > 1$. Notice that, if $\gcd(\lambda_1, \lambda_i) = d$, then $\{x\lambda_1 + y\lambda_i : x, y \in \mathbb{N}_0\} = d\langle \frac{\lambda_1}{d}, \frac{\lambda_i}{d}\rangle$, where, by $d\langle \frac{\lambda_1}{d}, \frac{\lambda_i}{d}\rangle$, we mean the set $\{d\lambda : \lambda \in \langle \frac{\lambda_1}{d}, \frac{\lambda_i}{d}\rangle\}$. Obviously, $d\langle \frac{\lambda_1}{d}, \frac{\lambda_i}{d}\rangle \subseteq \Lambda$. The next lemma is proved in [22].

Lemma 5. *If $\gcd(\lambda_1, \lambda_i) = d$, then $q(\lambda_i - \lambda_1) \in d\langle \frac{\lambda_1}{d}, \frac{\lambda_i}{d}\rangle$ if and only if $qd \leqslant \lfloor \frac{qd}{\lambda_1} \rfloor \lambda_i$. In particular, if $q \leqslant \lfloor \frac{q}{\lambda_1} \rfloor \lambda_i$, then $q(\lambda_i - \lambda_1) \in d\langle \frac{\lambda_1}{d}, \frac{\lambda_i}{d}\rangle$.*

Now, one can deduce the next result.

Proposition 1. *Suppose $\lambda_1 < \lambda_2 < \cdots < \lambda_n$ and let $\Lambda = \langle \lambda_1, \lambda_2, \ldots, \lambda_n \rangle$. If $q \leqslant \lfloor \frac{q}{\lambda_1} \rfloor \lambda_2$ then $GM_q(\Lambda) = L_q(\Lambda)$.*

Remark 3. *We have seen that for two-generated semigroups the converse is also true. For semigroups with any number of generators, the converse is not true in general. As a counterexample, let $\Lambda = \langle 5, 7, 18 \rangle = \{0, 5, 7, 10, 12, 14, 15, 17, 18, \ldots\}$ and consider $q = 9$. We have $\Lambda \setminus \cup_{\lambda_i \text{ generator of } \Lambda}(q\lambda_i + \Lambda) = \{0, 5, 7, 10, 12, 14, 15, 17, 18, 19, 20, 21, 22, 23, 24, 25, 26, 27, 28, 29, 30, 31, 32, 33, 34, 35, 36, 37, 38, 39, 40, 41, 42, 43, 44, 46, 47, 48, 49, 51, 53, 54, 56, 58, 61\} = \Lambda \setminus (q\lambda_1 + \Lambda)$. Hence, $GM_q(\langle 5, 7, 18\rangle) = 46$, which coincides with $L_q(\langle 5, 7, 18\rangle)$. Observe though that q, which is 9 is strictly larger than $\lfloor \frac{q}{\lambda_1} \rfloor \lambda_2$, which is 7.*

5.3. Simplified Computation

In [22], it was investigated whether the computation of the number $\Lambda \setminus \cup_{\lambda_i \text{ generator of } \Lambda}(q\lambda_i + \Lambda)$ could be performed as the simpler computation of $\Lambda \setminus \cup_{i \in J}(q\lambda_i + \Lambda)$ for some proper subset of indices $J \subseteq \{1, \ldots, n\}$. This is the purpose of the next lemma.

Lemma 6. *Suppose that $\Lambda = \langle \lambda_1, \ldots, \lambda_n \rangle$ and let $J \subseteq \{1, \ldots, n\}$ be an index subset. The following statements are equivalent*

1. $\Lambda \setminus \cup_{i=1}^{n}(q\lambda_i + \Lambda) = \Lambda \setminus \cup_{i \in J}(q\lambda_i + \Lambda)$;

2. For all $i \notin J$ there exists $1 \leqslant j \leqslant n$, $j \in J$ such that $q(\lambda_i - \lambda_j) \in \Lambda$.

The next lemma is a consequence of the previous one.

Lemma 7. *Suppose that $\Lambda = \langle \lambda_1, \ldots, \lambda_n \rangle$, where $\lambda_1 < \lambda_2 < \cdots < \lambda_n$ and suppose that $\lambda_1 < q$.*

1. *If λ_j is the maximum of the generators that are strictly smaller than $\frac{q}{\lfloor \frac{q}{\lambda_1} \rfloor}$, then $\Lambda \setminus \cup_{i=1}^{n}(q\lambda_i + \Lambda) = \Lambda \setminus \cup_{i=1}^{j}(q\lambda_i + \Lambda)$.*
2. *If λ_j is the maximum of the generators that are strictly smaller than $2\lambda_1 - 1$, then $\Lambda \setminus \cup_{i=1}^{n}(q\lambda_i + \Lambda) = \Lambda \setminus \cup_{i=1}^{j}(q\lambda_i + \Lambda)$.*

Proof. The first statement follows directly from Lemma 5 and Lemma 6. To prove the second statement, assume that $q = x\lambda_1 + y$, where $x \geq 1$ and y are integers. Then, $\frac{q}{\lfloor \frac{q}{\lambda_1} \rfloor} = \lambda_1 + \frac{y}{x}$ and the statement is a consequence of the inequalities $x \geqslant 1$ and $y \leqslant \lambda_1 - 1$. □

We call *Geil–Matsumoto generators* those generators that are strictly smaller than $2\lambda_1 - 1$. As follows from the previous results, to compute the Geil–Matsumoto bound, one only needs to subtract the ideals $q\mu + \Lambda$ from Λ for μ a Geil–Matsumoto generator. Because of the fact that, in general, one needs to subtract the ideals $q\lambda + \Lambda$ for *all* generators λ, this gives a computational improvement. In [22], we observed that, although it decreases with the genus, the portion of non-Geil–Matsumoto generators remains still significant for genus 25 with a portion of more than 30%.

We notice that Lemma 7 is a direct consequence of Lemma 6. We leave it as an open research problem to find other consequences of Lemma 6 to find further computational improvements.

6. Ideals and Isometry-Dual Sequences of One-Point Algebraic-Geometry Codes

6.1. Characterization of Isometry-Dual Sequences of Algebraic-Geometry Codes by Means of Sparse Ideals

We say that the codes $C, D \subseteq \mathbb{F}_q^n$ are isometric with respect to x, for $x \in \mathbb{F}_q^n$ if $D = \chi_x(C)$, where χ_x is the map $\chi_x : \mathbb{F}_q^n \to \mathbb{F}_q^n$ defined component-wise by $\chi_x(v) = x * v$. More generally, we say that the sequence $(C^{(i)})_{i=0,\ldots,n}$ of codes satisfies the *isometry-dual condition* if a vector $x \in (\mathbb{F}_q^*)^n$ exists so that $C^{(i)}$ is x-isometric to $\left(C^{(n-i)}\right)^{\perp}$ for every $i = 0, 1, \ldots, n$. Suppose now that $P_1, \ldots, P_n, Q$ are different rational points of a projective, smooth, irreducible curve of genus g and let $C_m = \{(f(P_1), \ldots, f(P_n)) : f \in L(mQ)\}$. As it has been previously stated, if Λ is the Weierstrass semigroup at Q and $\Lambda^* = \{0\} \cup \{m \in \mathbb{N}, m > 0 : C_m \neq C_{m-1}\} = \{m_0 = 0, m_1, \ldots, m_n\}$, then $\Lambda \setminus \Lambda^*$ is an ideal of Λ. Furthermore, Geil, Munuera, Ruano, and Torres proved the next lemma for $n > 2g + 2$ (in a different but equivalent formulation). The strict inequality was improved to a non-strict inequality in [24].

Lemma 8 (([17] Proposition 4.3.)). *Let $\Lambda^* = \{m_0, \ldots, m_n\}$ be as defined with $n \geq 2g + 2$. The sequence of codes $C_{m_0}, C_{m_1}, \ldots, C_{m_n}$ satisfies the isometry-dual poperty whenever $2g + n - 1 \in \Lambda^*$. Equivalently, the sequence $C_{m_0}, C_{m_1}, \ldots, C_{m_n}$ satisfies the isometry-dual property if and only if the ideal $\Lambda \setminus \Lambda^*$ is maximum sparse.*

6.2. Inclusion Relationship of Sparse Ideals

As seen in Theorem 2, a proper ideal I of Λ is maximum sparse if and only if I is of the form $\Lambda \setminus D(i)$ for some integer i satisfying $G(i) = 0$. The next lemma states the relationship between Frobenius numbers of maximum sparse ideals of a given numerical semigroup when the ideals satisfy inclusion relationships.

Lemma 9. *For two proper maximum sparse ideals I, I' of a numerical semigroup Λ with Frobenius numbers $\lambda_i, \lambda_{i'}$, the following statements are equivalent:*

1. $I \subseteq I'$;
2. $\Lambda \setminus I' \subseteq \Lambda \setminus I$;
3. $D(i') \subseteq D(i)$;
4. $\lambda_i - \lambda_{i'} \in \Lambda$;
5. $\#(\Lambda \setminus I) - \#(\Lambda \setminus I') \in \Lambda$.

Proof. The equivalence of statements (1) and (2) is obvious. Since I, I' are proper maximum sparse ideals, $D(i) = \Lambda \setminus I$ and $D(i') = \Lambda \setminus I'$. Hence, statement (2) and statement (3) are equivalent. Statement (3) is equivalent to $\lambda_{i'} \in D(i)$, which, in turn, is equivalent to statement (4). Statements (4) and (5) are equivalent since $\lambda_i = 2g - 1 + \#(\Lambda \setminus I)$ and $\lambda_{i'} = 2g - 1 + \#(\Lambda \setminus I')$. Hence, $\lambda_i - \lambda_{i'} = \#(\Lambda \setminus I) - \#(\Lambda \setminus I')$. □

6.3. Puncturing Sequences of Isometry-Dual One-Point Algebraic-Geometry Codes

We wonder now whether the isometry-dual property is inherited after puncturing sequences of one-point algebraic-geometry codes. We proved in the next theorem a necessary condition for the inheritance of the isometric-dual property. In particular, in order to maintain the property, the number of evaluating points that are supressed when puncturing must be a non-gap of the associated Weierstrass semigroup. This result was proved first in [24].

Theorem 4. *Suppose now that $P_1, \ldots, P_n, Q$ are different rational points of a projective, smooth, irreducible curve of genus g. Let Λ be the Weierstrass semigroup at Q, let $C_m = \{(f(P_1), \ldots, f(P_n)) : f \in L(mQ)\}$, and let $\Lambda^* = \{0\} \cup \{m \in \mathbb{N}, m > 0 : C_m \neq C_{m-1}\} = \{m_0 = 0, m_1, \ldots, m_n\}$. Suppose that the sequence $C_{m_0}, \ldots, C_{m_n}$ holds the isometry-dual property. Consider a subset $\{P_{i_1}, \ldots, P_{i_{n'}}\} \subseteq \{P_1, \ldots, P_n\}$, with $2g + 2 \leq n' < n$, the punctured codes $C'_m = \{(f(P_{i_1}), \ldots, f(P_{i_{n'}})) : f \in L(mQ)\}$, and the associated index set $(\Lambda^*)' = \{0\} \cup \{m \in \mathbb{N}, m > 0 : C'_m \neq C'_{m-1}\} = \{m'_1 = 0, m'_2, \ldots, m'_{n'}\}$. If the code sequence $\{0\}, C'_{m'_1}, C'_{m'_2}, \ldots, C'_{m'_{n'}}$ also holds the isometry-dual property, then $n - n' \in \Lambda$.*

Proof. By hypothesis, the set $\Lambda \setminus \Lambda^*$ is a maximum sparse ideal. If the sequence $\{0\}, C'_{m'_1}, \ldots, C'_{m'_{n'}}$ also holds the isometry-dual preperty, then so is $\Lambda \setminus (\Lambda^*)'$. We have $(\Lambda^*)' \subseteq \Lambda^*$ because $C'_m \neq C'_{m-1}$ implies $C_m \neq C_{m-1}$. Consequently, $\Lambda \setminus (\Lambda^*)' \supseteq \Lambda \setminus \Lambda^*$. Using Lemma 9, we can conclude that $\#\Lambda^* - \#(\Lambda^*)' = n - n' \in \Lambda$. □

7. Ideals and Generalized Hamming Weights

The number of nonzero coordinates of a word coincides with the cardinality of the support of the one-dimension vector it generates. Hence, the minimum distance of a linear code can be thought as the minimum number of elements the support of a one-dimension linear space can have. This is generalized to the so-called Hamming weights, which are defined, for each given dimension as the minimum size of the support of the linear subspaces of that dimension. The generalized Hamming weights for algebraic-geometry codes have been analyzed in [25–27]. Applications of generalized Hamming weights appear in a variety of fields of communications. Wei [28] first used the notion to analyze the performance of Ozarow–Wuyner's wire-tap channel of type II [29] and in connection to t-resilient functions. In [30], there is an update of the connections of generalized Hamming weights with the wire-tap channel using network coding. The reference [31] generalizes the notion for network coding. Generalized Hamming weights have applications also in the area of list decoding [32,33]. In particular, Guruswami showed that his (e, L)-list decodability notion in the case of erasures is equivalent to the generalized Hamming weights for linear codes. Generalized Hamming weights have also been used to bound the covering radius of linear codes [34] and for secure secret sharing based on linear codes [35,36]. One further related notion is that of relative generalized Hamming weights, where only the support of subspaces with no intersection with a given subspace are considered. They are applied to bound the information leakage in linear ramp secret sharing schemes. They were proposed in [37] and analyzed for algebraic-geometry codes in [38,39].

Heijnen and Pellikaan introduced in [40] the generalized order bounds for the generalized Hamming weights of dual one-point algebraic-geometry codes in terms of Weierstrass semigroups. Farrán and Munuera showed the existence of a constant, which they named the Feng–Rao number, depending only on the dimension of the Hamming weights and the Weierstrass semigroup, which completely determined the order bounds for codes of rate low enough. The references [41–44] deal with the generalized order bounds and the Feng–Rao numbers related to particular classes of semigroups.

We will present a new bound on the generalized Hamming weights that was first proved in [6]. It uses a lower bound on the Feng–Rao numbers derived from the upper bound for the Frobenius number of an ideal of a semigroup that we presented in Theorem 1. It is obtained through the analysis of intervals of consecutive gaps of Weierstrass semigroups. The idea of consecutive gaps was already used in [45] to bound the minimum distance of one-point codes and in [46] to bound the generalized Hamming weights for primal codes.

7.1. Feng–Rao Numbers

In Section 3.3, we introduced the ν sequence of a numerical semigroup Λ counting the number of pairs of non-gaps whose sum equals a given non-gap. The minimum distance of the dual one-point code $C_m^\perp$ associated with a rational point Q with Weierstrass semigroup Λ and associated sequence ν is bounded by the order (or Feng–Rao) bound defined as $\delta(m) = \min\{\nu_i : i > m\}$ [3,10,47]. Some results about the computation of the order bound can be found in [3,7,11–14,48].

The order bound for the minimum distance is generalized to any dimension r by the r-th order bound for the generalized r-th generalized Hamming weight. In this case, define $D(i_1, \ldots, i_r) = D(i_1) \cup \cdots \cup D(i_r)$. Then, the r-th order bound is defined as $\delta_r(m) = \min\{\#D(i_1, \ldots, i_r) : i_1, \ldots, i_r > m\}$. This definition was introduced in [40]. Farrán and Munuera proved in [49] that, for each integer $r \geq 2$ and for each numerical semigroup Λ, there exists a constant $E_r = E(\Lambda, r)$, the so-called r-th Feng–Rao number, satisfying that

1. $\delta_r(m) = m - g + E_r + 2$ for every m with $\lambda_m \geq 2c - 2$ ([49], Theorem 3),
2. $\delta_r(m) \geq m - g + E_r + 2$ for every m with $\lambda_m \geq c$ ([49], Theorem 8),

where g and c stand respectively for the genus and the conductor of Λ. This is indeed an extension of the Goppa bound in which case $r = 1$ and $E_r = 0$ ([3], Theorem 5.24). The constant E_r satisfies

3. $r \leq E_r \leq \lambda_{r-1}$ if $g > 0$ (and $r \geq 2$) ([49], Proposition 5),
4. $E_r = \lambda_{r-1}$ if $r \geq c$ ([49], Proposition 5),
5. $E_r = r - 1$ if $g = 0$.

In [41,49,50], one can find more results related to the Feng–Rao numbers.

We will use Theorem 1 to describe a new lower bound for the Feng–Rao number E_r. The new bound is strictly better than the bound $E_r \geq r$ for semigroups having more than two intervals of gaps and dimensions $r > 2$.

7.2. Bound on the Feng–Rao Numbers

To prove the new bound, we first need the next lemma, whose proof can be found in [6], and then we can state the theorem with the bound. The proof of the theorem uses that $\delta_r(m)$ counts the number of elements of a numerical semigroup not belonging to an ideal and the bound of Theorem 1.

Lemma 10. *Let*

$$\mathcal{A}(r, \ell, a_1, a_r) = \{A \subset \mathbb{N}_0 : \#A = r, \min(A) = a_1, \max(A) = a_r, A \text{ contains at least } \ell \text{ consecutive integers}\}.$$

For every $A \in \mathcal{A}$, let $\alpha(A) = \max\{a \in A : a+1-\ell, \ldots, a \in A\}$. Then, $\min \alpha(A) = \max\{a_1 + \ell - 1, a_1 + (\ell - 1)(a_1 - a_r) + \ell(r-1)\}$.

Theorem 5. *Suppose that $n_{\ell-1}$ is the number of intervals of at least $\ell - 1$ consecutive gaps of Λ, for ℓ an integer larger than 1. Then,*

$$E_r \geq \min\left\{r + \left\lceil \frac{r}{\ell-1} \right\rceil - 2, r + \left\lceil \frac{(\ell-1)n_{\ell-1}}{\ell} \right\rceil - 1\right\}. \tag{5}$$

Proof. By definition of $\delta_r(m)$, there exist integers $i_1, \ldots, i_r$ with $m < i_1 < \cdots < i_r$ such that $\delta_r(m) = \#D(i_1, \ldots, i_r)$. The integers $i_1, \ldots, i_r$ minimize $\#D(i_1, \ldots, i_r)$. Denote A the set $\{i_1, \ldots, i_r\}$. Suppose that the integer m is at least $2c - g - 1$. From the definition of E_r, we have $\delta_r(m) = m - g + E_r + 2$.

As the set A minimizes the amount $\#D(i_1, \ldots, i_r)$, then $i_1 = m + 1$. Now, one can apply Theorem 1 to the ideal $\Lambda \setminus D(i_1, \ldots, i_r)$, and obtain $(m - g + E_r + 2) + (2g - 1) \geq \lambda_{i_r} = g + i_r$. One can reorganize the inequality and obtain

$$i_r \leq m + E_r + 1. \tag{6}$$

If we assume that A has no ℓ consecutive integers, then

$$i_r \geq m + r + \left\lceil \frac{r - (\ell-1)}{\ell - 1} \right\rceil. \tag{7}$$

Then, by inequality (6), $E_r \geq r + \left\lceil \frac{r}{\ell-1} \right\rceil - 2$. On the other hand, assume that A has at least ℓ consecutive integers. Suppose that i_j is the maximum integer belonging to A so that $i_j - \ell + 1, \ldots, i_j \in A$ and so $i_{j-\ell+1} = i_j - \ell + 1, \ldots, i_{j-1} = i_j - 1$ and $\lambda_{i_{j-\ell+1}} = \lambda_{i_j} - \ell + 1, \ldots, \lambda_{i_{j-1}} = \lambda_{i_j} - 1$. Let $\Gamma = \{\lambda \in \Lambda : \lambda + 1, \ldots, \lambda + \ell - 1 \notin \Lambda\}$. In particular, if λ is an element of Γ, it must be strictly smaller than the conductor c of Λ. Obviously, $\#\Gamma = n_{\ell-1}$. If $\lambda \in \Gamma$, then $(\lambda_{i_j} - 1) - \lambda \in D(i_{j-1}) \setminus D(i_j), \ldots, (\lambda_{i_j} - \ell + 1) - \lambda \in D(i_{j-\ell+1}) \setminus D(i_j)$, and so $\{\lambda_{i_j} - \lambda - 1, \lambda_{i_j} - \lambda - 2, \ldots, \lambda_{i_j} - \lambda - \ell + 1\} \subseteq D(i_{j-\ell+1}, \ldots, i_{j-1}) \setminus D(i_j)$. In fact, $\cup_{\lambda \in \Gamma}\{\lambda_{i_j} - \lambda - 1, \ldots, \lambda_{i_j} - \lambda - \ell + 1\} \subseteq D(i_{j-\ell+1}, \ldots, i_{j-1}) \setminus D(i_j)$ and the sets in this union are disjoint. Indeed, for $\lambda, \lambda' \in \Gamma$, with $\lambda > \lambda'$, it holds $\lambda - \lambda' \geq \ell$. Then, $\min\{\lambda_{i_j} - \lambda' - 1, \ldots, \lambda_{i_j} - \lambda' - \ell + 1\} = \lambda_{i_j} - \lambda' - \ell + 1 \geq \lambda_{i_j} - \lambda + 1 > \max\{\lambda_{i_j} - \lambda - 1, \ldots, \lambda_{i_j} - \lambda - \ell + 1\}$. Hence,

$$\#D(i_1, \ldots, i_r) \geq \#D(i_{j-\ell+1}, \ldots, i_j) \geq (\ell-1)n_{\ell-1} + \nu_{i_j} = (\ell-1)n_{\ell-1} + i_j - g + 1 \tag{8}$$

Since $D(i_1, \ldots, i_r) = m - g + E_r + 2$, we get that $m - g + E_r + 2 \geq (\ell-1)n_{\ell-1} + i_j - g + 1$, so

$$E_r \geq (\ell-1)n_{\ell-1} + i_j - m - 1. \tag{9}$$

Now, by Lemma 10, and by the maximality of j,

$$i_j \geq \max\{i_1 + \ell - 1, i_1 + (\ell-1)(i_1 - i_r) + \ell(r-1)\}. \tag{10}$$

This implies

$$i_j \geq i_1 + \ell - 1, \tag{11}$$

and

$$i_j \geq i_1 + (\ell-1)(i_1 - i_r) + \ell(r-1). \tag{12}$$

On one side, we can use inequality (9) and inequality (11), and obtain $E_r \geq (\ell-1)(n_{\ell-1}+1)$.
On the other side, we can use inequality (9) and inequality (12), and then inequality (6), as follows:

$$\begin{aligned}
E_r &\geq (\ell-1)n_{\ell-1} + i_1 + (\ell-1)(i_1 - i_r) + \ell(r-1) - m - 1 \\
&= (\ell-1)n_{\ell-1} + (\ell-1)(i_1 - i_r) + \ell(r-1) \\
&\geq (\ell-1)n_{\ell-1} - (\ell-1)E_r + \ell(r-1),
\end{aligned}$$

from where we can conclude that $E_r \geq r - 1 + \left\lceil \frac{(\ell-1)n_{\ell-1}}{\ell} \right\rceil$.

At this point, we have shown that either $E_r \geq r + \left\lceil \frac{r}{\ell-1} \right\rceil - 2$ or $E_r \geq \max\{(\ell-1)(n_{\ell-1}+1), r + \left\lceil \frac{(\ell-1)n_{\ell-1}}{\ell} \right\rceil - 1\}$, depending on whether A has or does not have ℓ consecutive integers. Hence, we deduce the bounds that follow:

$$E_r \geq \min\{r + \left\lceil \frac{r}{\ell-1} \right\rceil - 2, (\ell-1)(n_{\ell-1}+1)\},$$

$$E_r \geq \min\{r + \left\lceil \frac{r}{\ell-1} \right\rceil - 2, r + \left\lceil \frac{(\ell-1)n_{\ell-1}}{\ell} \right\rceil - 1\}.$$

Let us see that the second bound is always at least as good as the first one. Hence, the first bound can be ignored. Indeed, if $r + \left\lceil \frac{r}{\ell-1} \right\rceil - 2 \leq r + \left\lceil \frac{(\ell-1)n_{\ell-1}}{\ell} \right\rceil - 1$, then we are done. Otherwise, if $r + \left\lceil \frac{r}{\ell-1} \right\rceil - 2 > r + \left\lceil \frac{(\ell-1)n_{\ell-1}}{\ell} \right\rceil - 1$, then we need to prove that $r + \left\lceil \frac{(\ell-1)n_{\ell-1}}{\ell} \right\rceil - 1 \geq (\ell-1)(n_{\ell-1}+1)$. If $r + \left\lceil \frac{r}{\ell-1} \right\rceil - 2 > r + \left\lceil \frac{(\ell-1)n_{\ell-1}}{\ell} \right\rceil - 1$, then $\left\lceil \frac{r}{\ell-1} \right\rceil > \left\lceil \frac{(\ell-1)n_{\ell-1}}{\ell} \right\rceil + 1$, which implies that $\frac{r}{\ell-1} > \frac{(\ell-1)n_{\ell-1}}{\ell} + 1$, and so $r > (\ell-1)(\frac{(\ell-1)n_{\ell-1}}{\ell} + 1) = (\ell-1)((n_{\ell-1}+1) - \frac{n_{\ell-1}}{\ell})$. This implies $r + \frac{(\ell-1)n_{\ell-1}}{\ell} > (\ell-1)(n_{\ell-1}+1)$, and so $r + \left\lceil \frac{(\ell-1)n_{\ell-1}}{\ell} \right\rceil - 1 \geq (\ell-1)(n_{\ell-1}+1)$, as desired. □

Remark 4. *The bound in Theorem 5 only improves the bound $E_r \geq r$ when $\ell < r/2 + 1$ and $n_{\ell-1} > 0$.*

7.3. Bound on the Generalized Hamming Weights

Corollary 2. *Let $\ell \geq 2$ and let m satisfy $\lambda_m \geq c$. Then, $\delta_r(m) \geq m + 2 - g + \min\left\{r - 2 + \left\lceil \frac{r}{\ell-1} \right\rceil, r - 1 + \left\lceil \frac{(\ell-1)n_{\ell-1}}{\ell} \right\rceil\right\}$.*

Remark 5. *From bound (5), taking $\ell = 2$, we deduce that, if n is the number of intervals of (at least one) gaps of Λ, then*

$$E_r \geq \min\{2(r-1), r + \lceil n/2 \rceil - 1\}. \tag{13}$$

Remark 6. *If $n \leq 2$ or $r = 2$, the bound in the previous remark equals the bound $E_r \geq r$. In any other case, this new bound is better.*

Corollary 3. *If the Weierstrass semigroup Λ has n intervals of gaps and its conductor is c, then, for every integer m such that $\lambda_m \geq c$,*

$$\delta_r(m) \geq \begin{cases} m - g + 2r, & \text{if } r \leq \lceil n/2 \rceil + 1, \\ m - g + r + \lceil n/2 \rceil + 1 & \text{otherwise.} \end{cases}$$

7.4. Sharpness of the Bound

If one analyzes the proof of Theorem 5, it can be seen that the bound (5) may only be sharp if

1. The inequality (6) is indeed an equality. That bound is obtained when one applies Theorem 1 to the ideal $\Lambda \setminus D(i_1, \ldots i_r)$. The inequality being an equality means applying Theorem 2 to the same ideal that $D(i_1, \ldots, i_r) = D(i_r)$. Hence, $i_1, \ldots, i_{r-1} \subseteq i_r - \Lambda$ and so, $i_r - i_{r-1} \geq \lambda_1$.

2. Either inequality (7) or both inequality (8) and inequality (10) are indeed equalities. In this case, $i_r - i_{r-1} \leq 2$.

From these observations, one can conclude that the bound may be sharp only if the Weierstrass semigroup Λ is a hyperelliptic semigroup, that is, a semigroup containing 2. For hyperelliptic semigroups, it was proved in ([50], Theorem 1) that $E_r = \lambda_{r-1} = 2(r-1)$. On the other hand, the bound (5) for the unique hyperelliptic semigroup of genus g is

$$E_r \geq \begin{cases} r-1, & \text{if } \ell > 2, \\ 2(r-1), & \text{if } \ell = 2 \text{ and } r-1 \leq \lceil g/2 \rceil, \\ r + \lceil g/2 \rceil - 1, & \text{if } \ell = 2 \text{ and } r-1 > \lceil g/2 \rceil. \end{cases}$$

Thus, we conclude that the bound is sharp if and only if $\ell = 2$, the Weierstrass semigroup Λ is hyperelliptic, and $r \leq \lceil g/2 \rceil + 1$.

7.5. The Bound Applied to the Hermitian Curve

The weight hierarchy of $\mathcal{H}_q$ has already been studied in [27,51]. However, for its simplicity, we wanted to give a description of n_ℓ. As we have seen before, the Weierstrass semigroup at the rational point at infinity is generated by q and $q+1$. Its weight hierarchy was studied in [42]. The semigroup generated by q and $q+1$ is $\{0\} \cup \{q, q+1\} \cup \{2q, 2q+1, 2q+2\} \cup \cdots \cup \{(q-2)q, \ldots, (q-2)q + (q-2)\} \cup \{k \in \mathbb{N}_0 : k \geq (q-1)q\}$. In this case, the lengths of the intervals of consecutive gaps are $q-1, q-2, \ldots, 1$. Thus,

$$n_\ell = \begin{cases} q - \ell, & \text{if } 1 \leq \ell \leq q, \\ 0, & \text{if } \ell \geq q. \end{cases}$$

It is left as an open question to compare the results in [41] with the bound proved in Theorem 5, using these values of n_ℓ.

8. Further Reading

It was our purpose to cite within the text the bibliography related to each specific section. However, the reader may be interested in some more general references. The books [52–54] have many results on numerical semigroups. Algebraic-geometry codes have been widely explained in different books such as [18,55,56] or in chapter [57]. For a general theory of one-point codes, their decoding, and also some of their relationships with Weierstrass semigroups, chapter [3] is probably the most important reference. Finally, chapter [1] is a survey of results on numerical semigroups, their classification, characterization and counting, and their relationship with algebraic-geometry codes from the perspective of decoding algorithms, their parameters such as the minimum distance, and the optimization of their redundancy under particular decoding restrictions.

9. Conclusions

Numerical semigroups play an important role in the analysis of error-correcting codes. More specifically, additive ideals of numerical semigroups are involved in determining non-redundant parity-checks, the code length, the generalized Hamming weights, and the isometry-dual sequences of algebraic-geometry codes. These results have been presented in this survey in a unified framework.

Funding: This work was partly supported by the Catalan Government under grant 2017 SGR 00705, by the Spanish Ministry of Economy and Competitivity under grant TIN2016-80250-R, and by Universitat Rovira i Virgili under grant OPEN2019.

Acknowledgments: The author would like to thank Michael E. O'Sullivan and Kwankyu Lee for many helpful discussions. She would also like to thank the coauthors of the main papers involved in this contribution: Kwankyu Lee, Albert Vico-Oton, Euijin Hong, and Iwan Duursma.

Conflicts of Interest: The author declares no conflict of interest.

References

1. Bras-Amorós, M. Numerical semigroups and codes. In *Algebraic Geometry Modeling in Information Theory*; Volume 8 of Ser. Coding Theory Cryptol.; World Science Publisher: Hackensack, NJ, USA, 2013; pp. 167–218.
2. Geil, O. On codes from norm-trace curves. *Finite Fields Appl.* **2003**, *9*, 351–371. [CrossRef]
3. Høholdt, T.; van Lint, J.H.; Pellikaan, R. Algebraic geometry codes. In *Handbook of Coding Theory*; North-Holland: Amsterdam, The Netherlands, 1998; Volumes I and II, pp. 871–961.
4. Sylvester, J.J. Mathematical questions with their solutions. *Educ. Times* **1884**, *41*, 21.
5. Stichtenoth, H. A note on Hermitian codes over $GF(q^2)$. *IEEE Trans. Inform. Theory* **1988**, *34*, 1345–1348. [CrossRef]
6. Bras-Amorós, M.; Lee, K.; Vico-Oton, A. New lower bounds on the generalized Hamming weights of AG codes. *IEEE Trans. Inform. Theory* **2014**, *60*, 5930–5937. [CrossRef]
7. Bras-Amorós, M. Acute semigroups, the order bound on the minimum distance, and the Feng-Rao improvements. *IEEE Trans. Inform. Theory* **2004**, *50*, 1282–1289. [CrossRef]
8. Bras-Amorós, M. A note on numerical semigroups. *IEEE Trans. Inform. Theory* **2007**, *53*, 821–823. [CrossRef]
9. Bras-Amorós, M.; O'Sullivan, M. On semigroups generated by two consecutive integers and improved Hermitian codes. *IEEE Trans. Inform. Theory* **2007**, *53*, 2560–2566. [CrossRef]
10. Kirfel, C.; Pellikaan, R. The minimum distance of codes in an array coming from telescopic semigroups. *IEEE Trans. Inform. Theory* **1995**, *41*, 1720–1732. [CrossRef]
11. Munuera, C.; Torres, F. A note on the order bound on the minimum distance of AG codes and acute semigroups. *Adv. Math. Commun.* **2008**, *2*, 175–181.
12. Oneto, A.; Tamone, G. On numerical semigroups and the order bound. *J. Pure Appl. Algebra* **2008**, *212*, 2271–2283. [CrossRef]
13. Oneto, A.; Tamone, G. On the order bound of one-point algebraic geometry codes. *J. Pure Appl. Algebra* **2009**, *213*, 1179–1191. [CrossRef]
14. Oneto, A.; Tamone, G. On some invariants in numerical semigroups and estimations of the order bound. *Semigroup Forum* **2010**, *81*, 483–509. [CrossRef]
15. Bras-Amorós, M. Addition behavior of a numerical semigroup. In *Arithmetic, Geometry and Coding Theory (AGCT 2003)*; Volume 11 of Sémin. Congr.; Société Mathématique de France: Paris, France, 2005; pp. 21–28.
16. Barucci, V. Decompositions of ideals into irreducible ideals in numerical semigroups. *J. Commut. Algebra* **2010**, *2*, 281–294. [CrossRef]
17. Geil, O.; Munuera, C.; Ruano, D.; Torres, F. On the order bounds for one-point AG codes. *Adv. Math. Commun.* **2011**, *5*, 489–504.
18. Stichtenoth, H. *Algebraic Function Fields and Codes*; Universitext; Springer: Berlin, Germany, 1993.
19. Geer, G.V.; Howe, E.W.; Lauter, K.E.; Ritzenthaler, C. Tables of Curves with Many Points. Available online: http://www.manypoints.org (accessed on 4 November 2019).
20. Lewittes, J. Places of degree one in function fields over finite fields. *J. Pure Appl. Algebra* **1990**, *69*, 177–183. [CrossRef]
21. Geil, O.; Matsumoto, R. Bounding the number of $\mathbb{F}_q$-rational places in algebraic function fields using Weierstrass semigroups. *J. Pure Appl. Algebra* **2009**, *213*, 1152–1156. [CrossRef]
22. Bras-Amorós, M.; Vico-Oton, A. On the Geil-Matsumoto bound and the length of AG codes. *Des. Codes Cryptogr.* **2014**, *70*, 117–125. [CrossRef]
23. Beelen, P.; Ruano, D. Bounding the number of points on a curve using a generalization of Weierstrass semigroups. *Des. Codes Cryptogr.* **2013**, *66*, 221–230. [CrossRef]
24. Bras-Amorós, M.; Duursma, I.; Hong, E. Isometry-dual flags of AG codes. 2019, submitted.
25. Munuera, C. On the generalized Hamming weights of geometric Goppa codes. *IEEE Trans. Inform. Theory* **1994**, *40*, 2092–2099. [CrossRef]
26. Munuera, C. Generalized Hamming weights and trellis complexity. In *Advances in Algebraic Geometry Codes*; Martinez-Moro, E., Munuera, C., Ruano, D., Eds.; World Scientific: Singapore, 2008; pp. 363–390.
27. Yang, K.; Kumar, P.V.; Stichtenoth, H. On the weight hierarchy of geometric Goppa codes. *IEEE Trans. Inform. Theory* **1994**, *40*, 913–920. [CrossRef]
28. Wei, V.K. Generalized Hamming weights for linear codes. *IEEE Trans. Inform. Theory* **1991**, *37*, 1412–1418. [CrossRef]

29. Ozarow, L.H.; Wyner, A.D. Wire-tap channel II. In *Advances in Cryptology (Paris, 1984)*; Volume 209 of Lecture Notes in Comput. Sci.; Springer: Berlin, Germany, 1985; pp. 33–50.
30. Rouayheb, S.E.; Soljanin, E.; Sprintson, A. Secure network coding for wiretap networks of type II. *IEEE Trans. Inform. Theory* **2012**, *58*, 1361–1371. [CrossRef]
31. Ngai, C.K.; Yeung, R.W.; Zhang, Z. Network generalized Hamming weight. *IEEE Trans. Inform. Theory* **2011**, *57*, 1136–1143. [CrossRef]
32. Gopalan, P.; Guruswami, V.; Raghavendra, P. List decoding tensor products and interleaved codes. In Proceedings of the 2009 ACM International Symposium on Theory of Computing, Bethesda, MD, USA, 31 May–2 June 2009; ACM: New York, NY, USA, 2009; pp. 13–22.
33. Guruswami, V. List decoding from erasures: bounds and code constructions. *IEEE Trans. Inform. Theory* **2003**, *49*, 2826–2833. [CrossRef]
34. Janwa, H.; Lal, A.K. On generalized Hamming weights and the covering radius of linear codes. In *Applied Algebra, Algebraic Algorithms and Error-Correcting Codes*; Volume 4851 of Lecture Notes in Comput. Sci.; Springer: Berlin, Germany, 2007; pp. 347–356.
35. Cruz, R.D.; Meyer, A.; Sole, P. An extension of Massey scheme for secret sharing. In Proceedings of the Information Theory Workshop, Dublin, Ireland, 30 August–3 September 2010.
36. Kurihara, J.; Uyematsu, T. Strongly-secure secret sharing based on linear codes can be characterized by generalized Hamming weight. In Proceedings of the 49th Annual Allerton Conference Communication, Control, and Computing, Monticello, IL, USA, 28–30 September 2011.
37. Luo, Y.; Mitrpant, C.; Vinck, A.J.H.; Chen, K. Some new characters on the wire-tap channel of type II. *IEEE Trans. Inform. Theory* **2005**, *51*, 1222–1229. [CrossRef]
38. Geil, O.; Martin, S.; Matsumoto, R.; Ruano, D.; Luo, Y. Relative generalized Hamming weights of one-point algebraic geometric codes. *IEEE Trans. Inform. Theory* **2014**, *60*, 5938–5949. [CrossRef]
39. Lee, K. Bounds for generalized Hamming weights of general AG codes. *Finite Fields Appl.* **2015**, *34*, 265–279. [CrossRef]
40. Heijnen, P.; Pellikaan, R. Generalized Hamming weights of q-ary Reed-Muller codes. *IEEE Trans. Inform. Theory* **1998**, *44*, 181–196. [CrossRef]
41. Delgado, M.; Farrán, J.I.; García-Sánchez, P.A.; Llena, D. On the generalized Feng-Rao numbers of numerical semigroups generated by intervals. *Math. Comp.* **2013**, *82*, 1813–1836. [CrossRef]
42. Delgado, M.; Farrán, J.I.; García-Sánchez, P.A.; Llena, D. On the weight hierarchy of codes coming from semigroups with two generators. *IEEE Trans. Inform. Theory* **2014**, *60*, 282–295. [CrossRef]
43. Farrán, J.I.; García-Sxaxnchez, P.A.; Heredia, B.A. On the second Feng-Rao distance of algebraic geometry codes related to Arf semigroups. *Des. Codes Cryptogr.* **2018**, *86*, 2893–2916. [CrossRef]
44. Farrán, J.I.; García-Sxaxnchez, P.A.; Heredia, B.A.; Leamer, M.J. The second Feng-Rao number for codes coming from telescopic semigroups. *Des. Codes Cryptogr.* **2018**, *86*, 1849–1864. [CrossRef]
45. García, A.; Kim, S.J.; Lax, R.F. Consecutive Weierstrass gaps and minimum distance of Goppa codes. *J. Pure Appl. Algebra* **1993**, *84*, 199–207. [CrossRef]
46. Tang, L. Consecutive Weierstrass gaps and weight hierarchy of geometric Goppa codes. *Algebra Colloq.* **1996**, *3*, 1–10.
47. Feng, G.L.; Rao, T.R.N. A simple approach for construction of algebraic-geometric codes from affine plane curves. *IEEE Trans. Inform. Theory* **1994**, *40*, 1003–1012. [CrossRef]
48. Campillo, A.; Farrán, J.I. Computing Weierstrass semigroups and the Feng-Rao distance from singular plane models. *Finite Fields Appl.* **2000**, *6*, 71–92. [CrossRef]
49. Farrán, J.I.; Munuera, C. Goppa-like bounds for the generalized Feng-Rao distances. *Discrete Appl. Math.* **2003**, *128*, 145–156. [CrossRef]
50. Farrán, J.I.; Sánchez, P.A.G.A.; Llena, D. On the Feng-Rao numbers. In *VII Jornadas de Matemática Discreta y Algorítmica*; CIEM: Castro Urdiales, Spain, 7–9 July 2010.
51. Barbero, A.I.; Munuera, C. The weight hierarchy of Hermitian codes. *SIAM J. Discrete Math.* **2000**, *13*, 79–104. [CrossRef]
52. Assi, A.; García-Sánchez, P.A. *Numerical Semigroups and Applications*; Volume 1 of RSME Springer Series; Springer: Cham, Switzerland, 2016.
53. Alfonsín, J.L.R. *The Diophantine Frobenius Problem*; Volume 30 of Oxford Lecture Series in Mathematics and its Applications; Oxford University Press: Oxford, UK, 2005.

54. Rosales, J.C.; García-Sánchez, P.A. *Numerical Semigroups*; Volume 20 of Developments in Mathematics; Springer: New York, NY, USA, 2009.
55. Pretzel, O. *Codes and Algebraic Curves*; Volume 8 of Oxford Lecture Series in Mathematics and its Applications; The Clarendon Press Oxford University Press: New York, NY, USA, 1998.
56. Van Lint, J.H.; van der Geer, G. *Introduction to Coding Theory and Algebraic Geometry*; Volume 12 of DMV Seminar; Birkhäuser: Basel, Switzerland, 1988.
57. Munuera, C.; Olaya-León, W. An introduction to algebraic geometry codes. In *Algebra for Secure and Reliable Communication Modeling*; Volume 642 of Contemp. Math.; American Mathematical Society: Providence, RI, USA, 2015; pp. 87–117.

© 2019 by the author. Licensee MDPI, Basel, Switzerland. This article is an open access article distributed under the terms and conditions of the Creative Commons Attribution (CC BY) license (http://creativecommons.org/licenses/by/4.0/).

Article
Algebraic Properties of the Block Cipher DESL

Kenneth Matheis [1,†], Rainer Steinwandt [2,†] and Adriana Suárez Corona [3,*,†]

1. Institute for Mathematics and Computer Science, Boca Raton, FL 33428, USA; kmatheis@imacs.org
2. Department of Mathematical Sciences, Florida Atlantic University, Boca Raton, FL 33431, USA; rsteinwa@fau.edu
3. Department of Mathematical Sciences, Universidad de León, 24071 León, Spain
* Correspondence: asuac@unileon.es
† These authors contributed equally to this work.

Received: 14 October 2019; Accepted: 12 November 2019; Published: 15 November 2019

Abstract: The Data Encryption Standard Lightweight extension (DESL) is a lightweight block cipher which is very similar to DES, but unlike DES uses only a single S-box. This work demonstrates that this block cipher satisfies comparable algebraic properties to DES—namely, the round functions of DESL generate the alternating group and both ciphers resist *multiple right-hand sides* attacks.

Keywords: lightweight cryptography; permutation group; block cipher

1. Introduction

Lightweight cryptography provides solutions tailored for devices with energy or computational constraints, which are increasingly present with the rapid increase of sensors and IoT devices. These requirements should not be met at the cost of losing security properties. Therefore, lightweight ciphers should ensure they offer similar security guarantees to their counterparts.

One of the protocols designed following these principles is DESL, a lightweight cipher very similar to the Data Encryption Standard (DES) [1], proposed by Leander et al. [2]. The proposed cipher introduces one radical change: all substitution boxes in the DES are replaced with a single new S-box. As detailed by Leander et al., this *DES Lightweight extension* (DESL) has very attractive features in terms of implementability on low-cost platforms. The obvious cryptanalytic question is whether these features might have been paid for with a loss of security. In other words, is the security of DESL comparable to that of the original DES? Leander et al.'s original paper [2] shows that DESL offers resistance against several common attack techniques, including certain types of linear and differential cryptanalyses. Finding structural weaknesses in DESL's design remains a challenge, so despite its short key length, DESL continues to attract interest and keeps getting cited [3–5]. Just a few days before submitting this manuscript, Ji et al. used DESL as a testing ground for proposed improvements of Matsui's algorithm [6]. In this contribution, we compare two algebraic properties of DESL with those of DES.

First we show that the round functions of DESL generate the same permutation group as the round functions of DES, namely the alternating group on 2^{64} points. Our proof strategy is the same as taken by Wernsdorf for DES [7], the core part being to establish 3-transitivity for the group in question. It is not surprising that the replacement of DES's S-boxes in DESL necessitates modifications of Wernsdorf's proof, and one might be tempted to hope that facing only one S-box (instead of several as in DES) simplifies the analysis—this did not seem to be the case for the S-box in question.

In the second part of the paper, we compare the resistance of full and reduced round versions of DES and DESL against an algebraic attack technique known as *multiple right-hand sides* (MRHS) [8]. This type of attack seems particularly interesting for Feistel ciphers like DES and DESL MRHS equations allow a fairly compact encoding of non-linear equations for the secret key, obtained from a known plaintext–ciphertext pair. The operations for solving such equations are in principle suitable for being accelerated through hardware [9], but establishing run-time estimates for such an attack against genuine ciphers is (perhaps unsurprisingly) challenging. While being devised as a tool for cryptanalysis, Raddum and Zajac recently demonstrated that a cipher representation derived from MRHS equations may yield a faster encryption than a reference implementation of a cipher [10]. In [11], Zajac leveraged MRHS equations as a tool to study the connection between the cost of algebraic attacks and the multiplicative complexity of lightweight ciphers. Here we consider the original cryptanalytic application of MRHS equations. The experimental results we found indicate that DESL offers resistance to this type of algebraic attack that is comparable to DES. As an aside, our results falsify a conjecture by Schoonen [12] (Hypothesis 5.1).

To keep our presentation reasonably self-contained, the next section presents the relevant details on the block cipher in question as well as the main ideas underlying an MRHS-based algebraic attack.

2. Preliminaries

With the exception of two modifications, DESL is identical to the Data Encryption Standard; in particular, plaintexts and ciphertexts are elements of $\{0,1\}^{64}$ and the key can be taken for an element of $\{0,1\}^{56}$. The first difference between DES and DESL is not relevant for the group-theoretic property and the algebraic attack we explore: unlike for DES, there is no initial permutation and no final permutation of the data processed in the cipher. The implications of the second modification is less obvious: DESL replaces all eight S-boxes in DES with a single new S-box.

2.1. Description of DESL

Figure 1 illustrates the basic data flow in DESL, and we refer to the DES specification [1] and Leander et al.'s paper [2] for a detailed specification. For our purposes it is enough to be aware of the following:

- There are 16 rounds, each round i implementing a permutation $\pi_i \in S_{2^{64}}$ which depends on a round key $K_i \in \{0,1\}^{48}$. The latter is derived from the secret key $K \in \{0,1\}^{56}$ through a suitable key schedule.
- Each of the 16 rounds involves a round-key-dependent function $F'_{K_i}(R_i) = P \circ \oplus \circ S \circ \oplus \circ E$ where
    - $E : \{0,1\}^{32} \longrightarrow \{0,1\}^{48}$ is an injective map specified in [1].
    - $\oplus : \{0,1\}^{48} \longrightarrow \{0,1\}^{48}, x \longmapsto x \oplus K_i$ adds (xor) the round key K_i to the input.
    - $S : \{0,1\}^{48} \longrightarrow \{0,1\}^{32}$ splits the input $(a_1, \ldots, a_{48}) \in \{0,1\}^{48}$ into 6-bit blocks and for each $j = 1, \ldots, 8$ substitutes $(a_{6j-5}, \ldots, a_{6j}) \in \{0,1\}^6$ with the corresponding 4-bit value obtained from Table 1.
    - $P \in S_{2^{32}}$ is a permutation on 32-bit strings as specified in [1].
- In each round, the 64-bit input is split into a left half $L_i \in \{0,1\}^{32}$ and a right half $R_i \in \{0,1\}^{32}$. Then the value $L'_i := F'_{K_i}(R_i) \oplus L_i$ is computed, where $\oplus$ is addition in $\{0,1\}^{48}$. The output of round i for $i \in \{1, \ldots, 15\}$ is (R_i, L'_i). In the last round there is no swap, that is, the value (L'_{16}, R_{16}) is output.

Table 1. The substitution function $S : \{0,1\}^6 \longrightarrow \{0,1\}^4$ of DESL is given by this S-box from [2]; $(a_1, \ldots, a_6) \in \{0,1\}^6$ is mapped to the 4-bit binary representation of the table entry in row no. $a_1 a_6$ and column no. $a_2 a_3 a_4 a_5$ (both interpreted as binary representation of a number in $\{0, \ldots, 3\}$ resp. $\{0, \ldots, 15\}$).

14	5	7	2	11	8	1	15	0	10	9	4	6	13	12	3
5	0	8	15	14	3	2	12	11	7	6	9	13	4	1	10
4	9	2	14	8	7	13	0	10	12	15	1	5	11	3	6
9	6	15	5	3	8	4	11	7	1	12	2	0	14	10	13

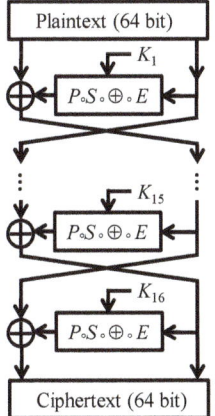

Figure 1. Data Encryption Standard Lightweight extension (DESL) overview.

For the group-theoretic part of our discussion of DESL, we make use of an observation about DES by Davio et al. [13] which has also been exploited in [7]. Namely, we rewrite DESL as shown in Figure 2, that is, by applying P^{-1} respectively P before the first round and after the last round, we combine E and P into a single function EP such that P no longer has to be applied after the application of the S-box. The composition of and E and P is given in Table 2.

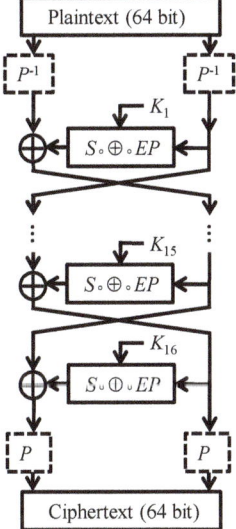

Figure 2. Equivalent description of DESL with the permutation P being applied before the expansion function E.

Table 2. The function $EP : \{0,1\}^{32} \longrightarrow \{0,1\}^{48}$, mapping $(a_1, \ldots, a_{32})$ to $a_{EP(1)}, \ldots, a_{EP(32)}$ where $EP(j)$ is the j-th entry in the table, reading from left to right, top to bottom (e.g., $EP(7) = 21$).

25	16	7	20	21	29
21	29	12	28	17	1
17	1	15	23	26	5
26	5	18	31	10	2
10	2	8	24	14	32
14	32	27	3	9	19
9	19	13	30	6	22
6	22	11	4	25	16

2.2. Multiple Right-Hand Sides (MRHS)

DESL, DES, and many other block ciphers can be modeled as series of polynomial equations over the binary field $\mathbb{F}_2$, therewith suggesting algebraic attacks as a possible attack vector. MRHS offers an alternative to algebraic attacks using SAT solvers or Gröbner bases. Instead of working with ordinary polynomials, equations are represented in a different way, which for several block ciphers, including DESL and DES, can be derived conveniently. For a detailed discussion of MRHS, we refer to Raddum and Semaev's work [8]. Here we restrict ourselves to an informal review of those aspects needed for our application. In particular, we do not discuss specifics of the implementation of the algorithm and refer to [8] (Section 6) for more details (cf. also [12,14]).

2.2.1. Basic Terminology

For a column vector $x = (x_1 \; x_2 \; \ldots \; x_y)^T \in \mathbb{F}_2^y$, a $k \times y$ binary matrix A of rank k, and column vectors $b_1, b_2, \ldots, b_s \in \mathbb{F}^k$ consider the following type of equation:

$$Ax = b_1, b_2, \ldots, b_s. \qquad (1)$$

We refer to such an equation as an *MRHS system of linear equations* with *right hand sides* $b_1, b_2, \ldots, b_s$. By a *solution* to (1) we mean a vector in $\mathbb{F}_2^y$ satisfying at least one particular linear system of equations $Ax = b_i$. The set of *all solutions* to (1) is obtained by forming the union of the solutions to the individual systems $Ax = b_i$ ($1 \leq i \leq s$). To work with MRHS systems of linear equations, we juxtapose the above column vectors b_i to form a matrix L and rewrite Equation (1) as $Ax = [L]$. The pair (A, L) is called a *symbol*, and when writing equations, the brackets around L emphasize that we are not working with an ordinary equation of matrices.

For example, the following is an MRHS system of linear equations:

$$\begin{pmatrix} 1 & 1 & 0 & 0 & 0 \\ 1 & 0 & 1 & 0 & 0 \\ 1 & 0 & 0 & 1 & 0 \end{pmatrix} \begin{pmatrix} x1 \\ x2 \\ x3 \\ x4 \\ x5 \end{pmatrix} = \begin{bmatrix} 1 & 0 & 0 & 1 \\ 0 & 1 & 0 & 0 \\ 0 & 0 & 1 & 1 \end{bmatrix}$$

and algebraically, it corresponds to the nonlinear equation

$$x_1 x_4 + x_1 x_2 + x_2 x_4 + x_2 + x_3 + x_4 + 1 = 0.$$

Given a system of symbols

$$\begin{aligned} S_1 : \quad A_1 x &= [L_1] \\ &\vdots \\ S_n : \quad A_n x &= [L_n] \end{aligned} \qquad (2)$$

a solution to such a system is defined in the obvious way: it is a vector $x \in \mathbb{F}_2^y$ satisfying all of the underlying n MRHS systems of linear equations, and the goal of the procedure discussed next is to identify all solutions of (2).

2.2.2. Solving a System of Symbols

There are three main components to MRHS: *agreeing*, *gluing*, and *extracting equations*. Since memory is finite in any actual implementation of the algorithm, it may also happen that we have to guess variables, and sometimes an equation symbol is made use of. Each of these parts is discussed below, and we start with a description of the main components.

Agreeing

The basic idea of an agreeing phase is to remove columns b in a right hand side L_i if no solution of $A_i x = b$ can be a solution to the system (2). To achieve this, pairwise *agreeing* of symbols is employed. Namely, let $S_i : A_i x = [L_i]$ and $S_j : A_j x = [L_j]$ be two symbols; we say that S_i and S_j agree if for every $b \in L_i$, there exists a $b' \in L_j$ such that the linear system

$$\begin{pmatrix} A_i \\ A_j \end{pmatrix} x = \begin{pmatrix} b \\ b' \end{pmatrix} \qquad (3)$$

is consistent, and, vice versa, for each $b' \in L_j$ there exists a $b \in L_i$ such that (3) is consistent.

In a situation where S_i and S_j do not agree, we remove those columns b from L_i for which the linear system $A_i x = b$ is inconsistent with $A_j x = [L_j]$. Dually, those columns b' from L_j are removed, for which $A_j x = b'$ is inconsistent with $A_i x = [L_i]$. Different strategies can be used to realize this basic idea, but for our purposes it is not necessary to go into further detail on this.

However, it is important to note that if two symbols S_h and S_i agree but S_i and S_j disagree, columns may be deleted in one or both of L_i and L_j. After this happens, it may well happen that S_h does not agree with either of the modified symbols, and it becomes necessary to *re-agree* S_h with them. During the latter agreement, columns from L_h may have to be deleted, and so on, possibly resulting in a chain reaction of column deletions. To ensure that a system of symbols reaches a pairwise-agreed state, we perform the *Agreeing1 algorithm* in Figure 3 (see [8] (Section 3.1)).

While the symbols in a System (2) do not pairwise agree,

1. Find S_i and S_j which do not agree.
2. Agree S_i and S_j.

Figure 3. Agreeing1 algorithm.

Gluing

When a system of symbols is in a pairwise-agreed state, we may choose to apply a different operation: The *gluing* of two symbols $S_i = (A_i, L_i)$ and $S_j = (A_j, L_j)$ results in a new symbol $Bx = [L]$ whose set of solutions is the set of common solutions to $A_i x = [L_i]$ and $A_j x = [L_j]$. After having formed this new symbol, it is inserted into the system at hand and the two symbols S_i and S_j which formed (B, L) are no longer necessary and are removed from the system.

Gluing a matrix L_i of width s_i with a matrix L_j of width s_j may yield a matrix L with as many as $s_i \cdot s_j$ columns. In an implementation, computing certain glues might therefore turn out to be infeasible, and one restricts to gluing only pairs of symbols where the number of columns in the resulting symbol does not exceed a certain threshold.

Once several glues have been performed, the symbols in the resulting system will usually no longer be pairwise-agreed, so the algorithm in Figure 3 can be run again, initiating another round of agreeing and gluing. The eventual goal of iterated agreeing and gluing steps is to obtain a system of symbols which consists of a single symbol.

Extracting Equations

From a given symbol $S : Ax = [L]$ we can try to extract *unique right-hand side (URHS)* equations, and if this is done, the resulting linear equations are placed in a dedicated symbol S_0 to which we refer as an *equation symbol*. The equation symbol is checked for consistency and size. The A-part of S_0 has the same number of columns as the A-parts of the other symbols, but its L-part has only one column. The equation symbol is not considered a proper part of the system (2) and does not take part in the Agreeing1 algorithm, nor is it removed after being glued to a symbol in the system. However, various implementations will involve S_0 in an agreement or gluing step. Furthermore, information from guessing variables may also be reflected by S_0.

Guessing Variables

It may happen that all symbols in a system are pairwise-agreed, no new URHS equations can be extracted, and no pair of symbols can be glued without exceeding the threshold. Lacking a better alternative, in such a situation one can guess the (one-bit) value of a variable. Before performing a guess, the system of symbols—to which we will refer as the *state*—is stored. After the guess has been made, pairwise agreeing, gluing, and equation extraction are performed as normal. If after some steps the state, again, does not allow for any new URHS equation to be computed or pair of symbols to be glued, the state is saved again, and we guess the value of another variable.

Obviously a guess for a variable can be incorrect, and this discovery manifests as follows: during the agreement of two symbols, all right-hand sides of at least one of the symbols get removed, indicating that the system has no solution. When this happens, the state can be rolled back to a previously saved state, so that a different guess can be made.

3. The Group Generated by DESL's Round Functions

In this section we show that the round functions of DESL generate the same group as the round functions of DES. The main part of the argument is to establish 3-transitivity of the group generated by DESL's round functions. To present the (somewhat technical) proof it will be convenient to introduce some notation.

3.1. Notation

The inputs for the S-box of DESL are bitstrings of length 6, outputting bit strings of length 4, as detailed in Table 1. The bitstring inputs are obtained by dividing a 48 bit string into eight blocks of equal length. To refer to the latter, given $a \in \{0,1\}^{48}$, we set $[a]_j := (a_i)_{i=6j-5}^{6j}$ $(j = 1, \ldots, 8)$. Analogously, for $a \in \{0,1\}^{32}$, we write $[a]_j := (a_i)_{i=4j-3}^{4j}$ $(j = 1, \ldots, 8)$ for the selection of 4-bit blocks. It will be clear from the context when we are dealing with 48-bit, respectively 32-bit values. Finally, as manifested in the balanced Feistel structure, splitting a bitstring of even length into two halves is a common operation in DESL, and for $(a_1, \ldots, a_{2m}) \in \{0,1\}^{2m}$ we define $a_L := (a_i)_{i=1}^{m} \in \{0,1\}^m$ and $a_R := (a_i)_{i=m+1}^{2m} \in \{0,1\}^m$.

Furthermore, for ease of readability, we will often represent bitstrings by the decimal number they represent in binary (again, the length of the bitstring will always be clear from the context). Accordingly, we write $A_{2^{64}}$ and $S_{2^{64}}$ for the alternating and symmetric group respectively on $\{0,1\}^{64}$. Given a set of permutations Π, we denote by $\langle \Pi \rangle$ the group generated by them. Specifically we are interested in the group G generated by the round functions F_K of DESL, where K ranges over all possible values in $\{0,1\}^{48}$. As in Wernsdorf's analysis of DES in [7], we ignore any restrictions imposed by the key schedule and allow the round keys to be chosen freely.

Using the description and notation from Section 2.1, for a given round key $K \in \{0,1\}^{48}$ we can represent $F_K \in S_{2^{64}}$ as

$$F_K : \begin{array}{ccc} \{0,1\}^{32} \times \{0,1\}^{32} & \longrightarrow & \{0,1\}^{32} \times \{0,1\}^{32} \\ (a,b) & \longmapsto & (b, ([a]_i \oplus S([K]_i \oplus [EP(b)]_i))_{i=1}^8) \end{array}.$$

We can therefore state our result in terms of these functions, proving that

$$G = \left\langle \{F_K \in S_{2^{64}} | K \in \{0,1\}^{48}\} \right\rangle = A_{2^{64}}.$$

3.2. Establishing 3-Transitivity of G

Before proving the main result, we will prove some previous lemmas.

Lemma 1. *The round functions of DESL generate a subgroup of $A_{2^{64}}$ that acts transitively on $\{0,1\}^{64}$.*

Proof. Verifying the transitivity of G is straightforward, and the work of Even and Goldreich [15] ensures that G is contained in the alternating group. □

As an intermediate step, we will show the transitivity of $G_0 := \{g \in G | g(0) = 0\}$ on $\{0,1\}^{64} \setminus \{(0,\ldots,0)\}$ and transitivity of $G_{0,d} := \{g \in G | g(0) = 0 \text{ and } g(d) = d\}$ on $\{0,1\}^{64} \setminus \{(0,\ldots,0),d\}$, where $d := (\delta_{31,i})_{i=1}^{64}$ has a single non-zero entry at the 31st position.

Before doing so, let us have a closer look at G_0 and $G_{0,d}$:

In view of the Feistel structure of DESL, it is perhaps not very surprising that we deal with pairs of round functions when exploring the transitivity of G_0 and $G_{0,d}$. We define four sets of key pairs, where the last two depend on the auxiliary value $d' := (0,0,0,1,0,0) \in \{0,1\}^6$:

$$\begin{aligned}
M &:= \{(k,k') \in \{0,1\}^6 \times \{0,1\}^6 | S(k) = S(k')\} \\
\mathbb{M} &:= \{(K,K') \in \{0,1\}^{48} \times \{0,1\}^{48} | \forall j \in \{1,\ldots,8\} : ([K]_j, [K']_j) \in M\} \\
M_{d'} &:= \{(k,k') \in M | S(k \oplus d') = S(k' \oplus d')\} \\
\mathbb{M}_{d'} &:= \{(K,K') \in \mathbb{M} | ([K]_4, [K']_4) \in M_{d'}.\}
\end{aligned}$$

The elements in G we are mainly interested in are of the form $F^L_{K,K'} := F_{K'}^{-1} F_K$ or $F^R_{K,K'} := F_{K'} F_K^{-1}$ with the key pair (K,K') being chosen from $\mathbb{M}$. For input pairs $(a,b) \in \{0,1\}^{32} \times \{0,1\}^{32}$ we have

$$\begin{aligned}
F^L_{K,K'}(a,b) &= ([a]_1 \oplus S([K]_1 \oplus [EP(b)]_1) \oplus S([K']_1 \oplus [EP(b)]_1),\ldots, \\
&\quad [a]_8 \oplus S([K]_8 \oplus [EP(b)]_8) \oplus S([K']_8 \oplus [EP(b)]_8), b) \text{ and} \\
F^R_{K,K'}(a,b) &= (a, [b]_1 \oplus S([K]_1 \oplus [EP(a)]_1) \oplus S([K']_1 \oplus [EP(a)]_1),\ldots, \\
&\quad [b]_8 \oplus S([K]_8 \oplus [EP(a)]_8) \oplus S([K']_8 \oplus [EP(a)]_8)).
\end{aligned}$$

In other words, when evaluating $F^L_{(K,K')}(a,b)$, the right half of the input does not vary and its left half is XORed with the value $(S([K]_i \oplus [EP(b)]_i) \oplus S([K']_i \oplus [EP(b)]_i))_{i=1}^8$ to the left half of the input. For $F^R_{(K,K')}$ the situation is similar, with the left half of the input being stabilized.

The following proposition helps in understanding the effect of repeatedly applying a map of the form $F^R_{K,K'}$, respectively $F^L_{K,K'}$.

Proposition 1. The functions $F_{K,K'}^L$ and $F_{K,K'}^R$ defined above satisfy the following:

(a) $\forall (K, K') \in \mathbb{M} : F_{K,K'}^L \in G_{0,d}$ and $F_{K,K'}^R \in G_0$.

(b) $\forall (K, K') \in \mathbb{M}_{d'} : F_{K,K'}^L \in G_{0,d}$ and $F_{K,K'}^R \in G_{0,d}$.

(c) Let $n \in \mathbb{N}$. Then, for all $(K_1, K_1'), \ldots, (K_n, K_n') \in \mathbb{M}$ and for all $(a, b) \in \{0,1\}^{32} \times \{0,1\}^{32}$, the following hold:

$$F_{K_1,K_1'}^R \circ \cdots \circ F_{K_n,K_n'}^R (a, b) =$$

$$\left(a, \quad [b]_1 \oplus \bigoplus_{i=1}^{n} (S([K_i]_1 \oplus [EP(a)]_1) \oplus S([K_i']_1 \oplus [EP(a)]_1)), \ldots, \right.$$

$$\left. [b]_8 \oplus \bigoplus_{i=1}^{n} (S([K_i]_8 \oplus [EP(a)]_8) \oplus S([K_i']_8 \oplus [EP(a)]_8)) \right)$$

and, analogously,

$$F_{K_1,K_1'}^L \circ \cdots \circ F_{K_n,K_n'}^L (a, b) =$$

$$\left([a]_1 \quad \oplus \quad \bigoplus_{i=1}^{n} (S([K_i]_1 \oplus [EP(b)]_1) \oplus S([K_i']_1 \oplus [EP(b)]_1)), \ldots, \right.$$

$$\left. [a]_8 \quad \oplus \quad \bigoplus_{i=1}^{n} (S([K_i]_8 \oplus [EP(b)]_8) \oplus S([K_i']_8 \oplus [EP(b)]_8)), \quad b \right).$$

Proof. The proof is immediate from the definition of $F_{K,K'}^L$ and $F_{K,K'}^R$. □

To understand better which values can be obtained in the left and right 32-bit halves of the output through repeated application of a map of the form $F_{K,K'}^R$ (respectively $F_{K,K'}^L$), given some 64-bit input, it is helpful to take a look at some $\mathbb{F}_2$-vector subspaces of $\mathbb{F}_2^4$:

Lemma 2. For $y \in \{0,1\}^6 \setminus \{(0,0,0,0,0,0)\}$ let

$$U(y) := \langle S(k \oplus y) \oplus S(k' \oplus y) \,|\, (k, k') \in \mathbb{M} \rangle \subseteq \mathbb{F}_2^4$$

be the $\mathbb{F}_2$-vector space spanned by $\{S(k \oplus y) \oplus S(k' \oplus y) \,|\, (k, k') \in \mathbb{M}\}$.
Similarly, denote by $U_{d'}(y)$ the $\mathbb{F}_2$-vector space

$$U_{d'}(y) := \langle S(k \oplus y) \oplus S(k' \oplus y) \,|\, (k, k') \in \mathbb{M}_{d'} \rangle.$$

Then, the following statements hold:

(a) $\forall y \in \{0,1\}^6 \setminus \{(0,0,0,0,0,0), (0,0,0,0,0,1)\} : U(y) = \{0,1\}^4$.

(b) $U(0,0,0,0,0,1) = \{0, 2, 4, 6, 8, 10, 12, 14\}$.

(c) $\forall y \in \{2, 6, 17, 18, 21, 22, 41, 45, 49, 53, 58, 62\} : U_{d'}(y) = \{0,1\}^4$.

(d) $\forall y \in \{0,1\}^6 \setminus \{(0,0,0,1,0,0)\} : U_{d'}(y) \neq \{0\}$.

Proof. The proof is by direct computation, e.g., using a programming language like Python [16]. □

Remark 1. *Bringing the notation in Lemma 2 to use, from Proposition 1 we obtain the following statements which for the case $U([EP(a)]_i) = \{0,1\}^4$ (respectively $U([EP(b)]_k) = \{0,1\}^4$) may be regarded as "hinting at transitivity":*

- For $i = 1, \ldots, 8$ let $u_i \in U([EP(a)]_i)$ be a bitstring. Then, there exist $(K_1, K'_1), \ldots, (K_n, K'_n) \in \mathbb{M}$ such that $F^R_{K_1, K'_1} \circ \cdots \circ F^R_{K_n, K'_n}(a, b) = (a, [b]_1 \oplus u_1, \ldots, [b]_8 \oplus u_8)$ for all $(a, b) \in \{0, 1\}^{32} \times \{0, 1\}^{32}$.

- For $i = 1, \ldots, 8$ let $u_i \in U([EP(b)]_i)$ be a bitstring. Then, there exist $(K_1, K'_1), \ldots, (K_n, K'_n) \in \mathbb{M}$ such that $F^L_{K_1, K'_1} \circ \cdots \circ F^L_{K_n, K'_n}(a, b) = ([a]_1 \oplus u_1, \ldots, [a]_8 \oplus u_8, b)$ for all $(a, b) \in \{0, 1\}^{32} \times \{0, 1\}^{32}$.

- For $i \in \{1, \ldots, 8\} \setminus \{4\}$ let $u_i \in U([EP(a)]_i)$ be a bitstring and let $u_4 \in U_{d'}([EP(a)]_4)$. Then, there exist $(K_1, K'_1), \ldots, (K_n, K'_n) \in \mathbb{M}_{d'}$ such that $F^R_{K_1, K'_1} \circ \cdots \circ F^R_{K_n, K'_n}(a, b) = (a, b_1 \oplus u_1, \ldots, b_8 \oplus u_8)$ for all $(a, b) \in \{0, 1\}^{32} \times \{0, 1\}^{32}$.

- For $i \in \{1, \ldots, 8\} \setminus \{4\}$ let $u_i \in U([EP(b)]_i)$ be a bitstring and let $u_4 \in U_{d'}([EP(b)]_4)$. Then there exist $(K_1, K'_1), \ldots, (K_n, K'_n) \in \mathbb{M}_{d'}$ such that $F^L_{K_1, K'_1} \circ \cdots \circ F^L_{K_n, K'_n}(a, b) = (a_1 \oplus u_1, \ldots, a_8 \oplus u_8, b)$ for all $(a, b) \in \{0, 1\}^{32} \times \{0, 1\}^{32}$.

Therefore, if we know that the equality $U([EP(a)]_k) = \{0, 1\}^4$ holds for some $1 \leq k \leq 8$, then for each bitstring $c \in \{0, 1\}^4$ we can find a sequence of key pairs $(K_1, K'_1), \ldots, (K_n, K'_n) \in \mathbb{M}$ with

$$\left[\left[F^R_{K_1, K'_1} \circ \cdots \circ F^R_{K_n, K'_n}(a, b)\right]_R\right]_k = c.$$

For instance, we can choose pairs $(K_1, K'_1), \ldots, (K_n, K'_n)$ with $([K_j]_k, [K'_j]_k) \in M$ corresponding to the linear combination of $c \oplus [b]_k$, and the rest of the positions being 0. This ensures that all (K_j, K'_j) are contained in $\mathbb{M}$, and if $U_{d'}([EP(a)]_k) = \{0, 1\}^4$ or $k \neq 4$, we can also ensure $(K_1, K'_1), \ldots, (K_n, K'_n) \in \mathbb{M}_{d'}$.

Similarly, in case $U([EP(b)]_k)$ contains all bitstrings of length 4, we can obtain a sequence of key pairs with

$$\left[\left[F^L_{K_1, K'_1} \circ \cdots \circ F^L_{K_n, K'_n}(a, b)\right]_L\right]_k = c.$$

The subsequent lemmata enable us to argue that $G_{0,d}$ acts transitively on $\{0, 1\}^{64} \setminus \{0, d\}$. In other words, we prove that for all $x, y \in \{0, 1\}^{64} \setminus \{0, d\}$ the equivalence $x \sim y$ holds, where $x \sim y \iff \exists g \in G_{0,d} : g(x) = y$. The proofs exploit in particular the transitivity of $\sim$.

Lemma 3. *Let $e := (1, 0, 1, \ldots, 1) \in \{0, 1\}^{32}$ be the 32-bit vector which has a single 0-entry at the second position and 1-entries everywhere else, and let $(z, z') \in \{0, 1\}^{32} \times \{0, 1\}^{32}$ be arbitrary. Then $(e, z) \sim (e, z')$.*

Proof. Let $(z, z') \in \{0, 1\}^{32} \times \{0, 1\}^{32}$ be arbitrary, but fixed. From Table 2 we see that

$$[EP(e)]_i = \begin{cases} (1, 1, 1, 1, 1, 1) & , \text{if } i \in \{1, 2, 3, 6, 7, 8\} \\ (1, 1, 1, 1, 1, 0) & , \text{if } i = 4 \\ (0, 1, 1, 1, 1, 1) & , \text{if } i = 5 \end{cases}$$

Hence, by properties (a) and (c) of Lemma 2 we obtain $U((EP(e))_i) = \{0, 1\}^4$ for all $i = 1, \ldots, 8$ as well as $U_{d'}((EP(e))_4) = \{0, 1\}^4$.

Therefore, because of Remark 1 for $c = (z'_1, z'_2, z'_3, z'_4)$ we get:

$(e, z) \sim (e, (z'_1, z'_2, z'_3, z'_4, z_5, \ldots, z_{32}))$, since $(e, (z'_1, z'_2, z'_3, z'_4, z_5, \ldots, z_{32})) = F^R_{K^1, K^{1'}} \circ \cdots \circ F^R_{K^n, K^{n'}}(e, z)$, for the corresponding $(K^i, K^{i'}), i \in \{1, \ldots, n\}$.

Analogously, since $U((EPe)_2) = \{0, 1\}^4$, we can obtain:

$(e, (z'_1, z'_2, z'_3, z'_4, z_5, \ldots, z_{32})) \sim (e, (z'_1, \ldots, z'_8, z_9, \ldots, z_{32}))$.

If we continue carrying out the same procedure, since all the subspaces considered are $\{0, 1\}^4$, we can finally see that $(e, z) \sim (e, z')$. □

Lemma 4. $\forall\, a \in \{0,1\}^{64} \setminus \{0,d\}$, $\exists\, a' \in \{0,1\}^{64} \setminus \{0,d\} : a' \sim a$ and $\exists\, i \in \{1,\ldots,32\} \setminus \{2,5,10,18,26,31\} : a'_i = 1$.

Proof. If $\exists\, i \in \{1,\ldots,32\} \setminus \{2,5,10,18,26,31\} : a_i = 1$, then we obtain the lemma with $a' := a$.

Otherwise, we distinguish two cases:

- If $\exists\, i \in \{33,\ldots,64\} : a_i = 1$:

  Then $\exists\, l \in \{1,\ldots,8\}$ such that $[EP(a)_{i=33}^{64}]_l \neq 0$:

  - If $[EP(a)_{i=33}^{64}]_l \neq 1$, then $U([EP(a)_{i=33}^{64}]_l) = \{0,1\}^4$. Therefore, because of Remark 1, we can show $a' = F^L_{K^1,K^{1'}} \circ \cdots \circ F^L_{K^n,K^{n'}}(a)$ such that $([a']_L)_j = 1$ for $j \in \{4l-3,\ldots,4l\}$. Thus, $\exists\, i \in \{1,\ldots,32\} \setminus \{2,5,10,18,26,31\} : a'_i = 1$.
  - If $[EP(a)_{i=33}^{64}]_l = 1$, then $U([EP(a)_{i=33}^{64}]_l) = \{0,2,4,6,8,10,12,14\}$. With an argument similar to the previous one, we can get an element $a' = F^L_{K^1,K^{1'}} \circ \cdots \circ F^L_{K^n,K^{n'}}(a)$, such that $(a'_L)_i = 1$ for $i \in \{4l-3,\ldots,4l-1\}$. Therefore, $\exists\, i \in \{1,\ldots,32\} \setminus \{2,5,10,18,26,31\} : a'_i = 1$.

- If $\forall\, i \in \{33,\ldots,64\} : a_i = 0$.

  Since $a \neq 0$, then $\exists\, i \in \{1,\ldots,32\} : a_i = 1$. Therefore, $\exists\, l \in \{1,\ldots,8\}$ such that $[EP(a)_{i=1}^{32}]_l \neq 0$ and, like before (but using "right-functions") we prove that we can get an element $a' = F^R_{K^1,K^{1'}} \circ \cdots \circ F^R_{K^n,K^{n'}}(a)$, where $(K^i, K^{i'}) \in \mathbb{M}_{d'}$, such that $\exists\, i \in \{33,\ldots,64\} : a'_i = 1$. Notice that in this case the pairs $(K^i, K^{i'})$ must be not only in $\mathbb{M}$, but in $\mathbb{M}_{d'}$, so that $a \sim a'$ (Proposition 1(b)).

  - If $l \neq 4$
    * If $(EP(a)_{i=1}^{32})_l \neq 1$, then $U([EP(a)_{i=1}^{32})]_l) = \{0,1\}^4$. Therefore, because of Remark 1, we can have $a' = F^R_{K^1,K^{1'}} \circ \cdots \circ F^R_{K^n,K^{n'}}(a)$, where $(K^i, K^{i'}) \in \mathbb{M}_{d'}$, with $a'_i = 1$ for some $i \in \{33,\ldots,64\}$.
    * If $[EP(a)_{i=1}^{32}]_l = 1$, then $U([EP(a)_{i=1}^{32})]_l) = \{0,2,4,6,8,10,12,14\}$. With the same argument as before, we can get an element $a' = F^R_{K^1,K^{1'}} \circ \cdots \circ F^R_{K^n,K^{n'}}(a)$, such that $a'_i = 1$ for $i = 32 + j$, where $j \in \{4l-3,\ldots,4l-1\}$.
  - If $l = 4$: Since $a \neq d$, according to Table 2, $(EPa)_4 \neq (0,0,0,1,0,0)$. Therefore, we have $U_{d'}((EPa)_4) \neq 0$ (Lemma 2(d)) and we can obtain, as in the previous cases, an element $a' := F^L_{K^1,K^{1'}} \circ \cdots \circ F^R_{K^n,K^{n'}}(a) \sim a$, with $a'_i = 1$ for some $i \in \{33,\ldots,64\}$.

Hence, this case is traced back to the case $\exists\, i \in \{33,\ldots,64\} : a_i = 1$ and the proof is complete. □

Lemma 5. $\forall\, a' \in \{0,1\}^{64} \setminus \{0,d\} : a' \sim a$ and $\exists\, i \in \{1,\ldots,32\} \setminus \{2,5,10,18,26,31\} : a'_i = 1$, $\exists\, a'' \in \{0,1\}^{64} \setminus \{0,d\} : a'' \sim a'$ and $\forall\, i \in \{1,\ldots,32\} \setminus \{13,\ldots,16\} : a''_i = e_i$.

Proof. If $\forall\, i \in \{1,\ldots,32\} \setminus \{13,\ldots,16\} : a''_i = e_i$, then we immediately obtain the Lemma with $a'' := a'$.

Otherwise, we choose an index $j \in \{1,\ldots,32\} \setminus \{2,5,10,18,26,31\} : a'_j = 1$ and we will prove that $\exists\, a^0 \in \{0,1\}^{64} \setminus \{0,d\} : a^0 \sim a'$, $[a^0]_L = [a']_L$ and $\forall\, i \in I(j) : (a^0)_{32+i} = 1$, where the sets $I(j)$ are defined in Figure 4.

j	$I(j)$
1	$\{5,\ldots,12\} \setminus \{8\}$
3 or 27	$\{21,\ldots,24\}$
4 or 11	$\{29,\ldots,32\}$
6	$\{25,\ldots,32\}$
7 or 20	$\{1,\ldots,4\}$
8 or 24	$\{17,\ldots,20\}$
9	$\{21,\ldots,29\}$
12 or 28	$\{5,\ldots,8\}$
13 or 30	$\{25,\ldots,28\}$
14	$\{17,\ldots,24\}$
15 or 23	$\{9,\ldots,12\}$
16	$\{1,\ldots,4\} \cup \{29,\ldots,31\}$
17	$\{5,\ldots,12\}$
19	$\{21,\ldots,28\} \setminus \{24\}$
21	$\{1,\ldots,8\}$
22	$\{25,\ldots,32\} \setminus \{28\}$
25	$\{1,\ldots,4\} \cup \{29,\ldots,32\}$
29	$\{1,\ldots,8\} \setminus \{4\}$
32	$\{17,\ldots,24\} \setminus \{20\}$

Figure 4. Definition of $I(j)$.

We define $a^0 := F^R_{K^1,K^{1'}} \circ F^R_{K^2,K^{2'}} \circ \cdots \circ F^R_{K^n,K^{n'}}(a')$, with $(K^i, K^{i'}) \in \mathbb{M}_{d'}$. Therefore, $[a^0]_L = [a']_L$, and we will see that if $(K^i, K^{i'}), i \in \{1,\ldots,n\}$ have been chosen appropriately, we can have $(a^0)_{32+i} = 1, \forall i \in I(j)$.

For $j = 1$:

According to Table 2, $[EP(a')_L]_2 \neq 0$ and $[EP(a')_L]_3 \notin \{0,1\}$, since the corresponding positions for a'_1 are 12 and 14, which are in blocks 2 and 3. Therefore, we have:

- If $[EP(a')_L]_2 \neq 1$, then $U([EP(a')_L]_2) = \{0,1\}^4$. Hence, because of Remark 1, $\exists\, (K^i, K^{i'}) \in \mathbb{M}_{d'}$ such that $[[a^0]_R]_2 = [F^L_{K^1,K^{1'}} \circ F^L_{K^2,K^{2'}} \circ \cdots \circ F^L_{K^n,K^{n'}}(a')]_2 = (1,1,1,1)$. Therefore, $(a^0)_{32+i} = 1$ for all $i \in \{5,\ldots,8\}$.

- If $[EP(a')_L]_2 = 1$, then $U([EP(a')_L]_2) = \{0,2,4,6,8,10,12,14\}$. With a similar argument, $\exists\, (K^i, K^{i'}) \in \mathbb{M}_{d'}$ such that $[[a^0]_R]_2 = [F^L_{K^1,K^{1'}} \circ F^L_{K^2,K^{2'}} \circ \cdots \circ F^L_{K^n,K^{n'}}(a')]_2 = (1,1,1,0)$. Therefore, $(a^0)_{32+i} = 1$ for all $i \in \{5,\ldots,7\}$.

Since $[EP(a')_L]_3 \notin \{0,1\}$, then $U([EP(a')_L]_3) = \{0,1\}^4$ and therefore $\exists (K^i, K^{i'}) \in \mathbb{M}_{d'}$ such that $[[a^0]_R]_3 = [F^L_{K^1,K^{1'}} \circ F^L_{K^2,K^{2'}} \circ \cdots \circ F^L_{K^n,K^{n'}}(a')]_3 = (1,1,1,1)$. Therefore, $(a^0)_{32+i} = 1$ for all $i \in \{9,\ldots,12\}$.

Thus, considering the composition of the functions involved, we obtain a^0 such that $(a^0)_{32+i} = 1, \forall i \in \{5,\ldots,12\} \setminus \{8\}$.

A similar argument applies to the other values of $j \in \{1,\ldots,32\} \setminus \{13,\ldots,16\}$.

Now, we will see that $\exists\, a^1 \in \{0,1\}^{64} \setminus \{0,d\} : a^1 \sim a^0$, $[a^1]_R = [a^0]_R$ and $\forall\, i \in J(j) : (a^1)_i = e_i$, where the sets $J(j)$ are defined in Figure 5.

We define $a^1 := F^L_{K^1,K^{1'}} \circ \cdots \circ F^L_{K^n,K^{n'}}(a')$, with $(K^i, K^{i'}) \in \mathbb{M}$. Therefore, $[a^0]_R = [a']_R$, and we will see that choosing adequate elements $(K^i, K^{i'})$, we can have $(a^1)_i = e_i, \forall\, i \in J(j)$.

For $j = 1, I(1) = \{5,\ldots,12\} \setminus \{8\}$:

According to Table 2, let us see which positions $EP(([a^0]_R)_i)$ are in for the different values of $i \in I(1)$. We can see $EP(([a^0]_R)_5)$ is in position 18 (block 3) and 20 (block 4), $EP(([a^0]_R)_6)$ is in position 41 (block 7) and 43 (block 8), $EP(([a^0]_R)_7)$ is in position 3 (block 1), $EP(([a^0]_R)_9)$ is in position 35 and 37 (blocks 6 and 7), $EP(([a^0]_R)_{10})$ is in position 23 and 25 (block 4 and 5), $EP(([a^0]_R)_{11})$ is in position 45 (block 8), and $EP(([a^0]_R)_{12})$ is in position 9 (block 2).

j	$(\{1,\ldots,32\} \setminus \{13,\ldots,16\}) \setminus J(j)$
1 or 17	$\{12\}$
3 or 27	$\{21,\ldots,24\} \cup \{28\}$
4 or 11	$\{4\} \cup \{9,\ldots,12\} \cup \{20\} \cup \{29,\ldots,32\}$
6 or 22	$\{20\}$
7 or 20	$\{1,\ldots,4\} \cup \{8\} \cup \{25,\ldots,28\}$
8 or 24	$\{17,\ldots,20\} \cup \{24\} \cup \{29,\ldots,32\}$
9	$\{28\}$
12 or 28	$\{5,\ldots,8\} \cup \{12\} \cup \{21,\ldots,24\}$
13 or 30	$\{17,\ldots,20\} \cup \{25,\ldots,28\}$
14 or 32	$\{24\}$
15 or 23	$\{1,\ldots,4\} \cup \{9,\ldots,12\}$
16 or 25	$\{4\}$
19	$\{17,\ldots,20\} \cup \{28\}$
21 or 29	$\{8\}$

Figure 5. Definition of $J(j)$.

In all blocks j, for $j \in \{1,\ldots,8\} \setminus \{3\}$, we have $[EP(a^0)_R]_j \notin \{0,1\}$ and then $U([EP(a^0)_R]_j) = \{0,1\}^4$. Therefore, as discussed in the previous proofs, $\exists (K^i, K^{i'}) \in \mathbb{M}$ such that $[[a^1]_L]_j := [F^L_{K^1,K^{1'}} \circ F^L_{K^2,K^{2'}} \circ \cdots \circ F^L_{K^n,K^{n'}}(a')]_j = [e]_j \; \forall j \in \{1,\ldots,8\} \setminus \{3\}$. For block 3, we have $[EP(a^0)_R]_3 = 1$, therefore $\exists (K^i, K^{i'}) \in \mathbb{M}$ such that $(a^1)_i := (F^L_{K^1,K^{1'}} \circ F^L_{K^2,K^{2'}} \circ \cdots \circ F^L_{K^n,K^{n'}}(a'))_i = e_i \; \forall i \in \{9,\ldots,11\}$.

Therefore, the only position we cannot assure is equal to e is $i = 12$, therefore $J(1)^c = \{12\}$.

For the rest of the indices j, we use similar arguments to compute sets $J(j)$.

- If $j \in \{1,6,9,14,16,17,21,22,25,29,32\}$, the set $(\{1,\ldots,32\} \setminus \{13,\ldots,16\}) \setminus J(j)$ has only one element. Therefore, as $((a^1)_L)_i = e_i \; \forall i \in J(j)$, $[EP(a^1_L)]_i \notin \{0,1\} \; \forall i \in \{1,\ldots,8\} \setminus \{4\}$, so $U([EP(a^1_L)]_i) = \{0,1\}^4$. Therefore, choosing appropriate $(K^i, K^{i'}) \in \mathbb{M}_{d'}$ we get $a^2 := F^R_{K^1,K^{1'}} \circ \cdots \circ F^R_{K^n,K^{n'}}(a^1)$, such that $([a^2]_R)_i = e_i \; \forall i \in \{1,\ldots,32\} \setminus \{13,\ldots,16\}$ (Remark 1).

  Therefore, we have $[EP(a^2_R)]_i \notin \{0,1\} \; \forall i \in \{1,\ldots,8\} \setminus \{4\}$, so $U([EP(a^2_R)]_i) = \{0,1\}^4$. Now, choosing adequate $(K^i, K^{i'}) \in \mathbb{M}_{d'}$, we can have $a^3 := F^L_{K^1,K^{1'}} \circ \cdots \circ F^L_{K^n,K^{n'}}(a^2)$, such that $(a^3)_i = e_i \; \forall i \in \{1,\ldots,32\} \setminus \{13,\ldots,16\}$. Therefore, for $a'' := a^3$ we have the desired result.

  Hence, we have seen that the lemma holds if $a'_i = 1$ for $j \in \{1,6,9,14,16,17,21,22,25,29,32\}$.

- For indices $j \in \{1,\ldots,32\} \setminus \{2,5,10,18,26,31\}$, we have $J(j) \cap \{1,6,9,14,16,17,21,22,25,29,32\} \neq \emptyset$. Therefore, we are in the case where $\exists j \in \{1,6,9,14,16,17,21,22,25,29,32\}$ such that $(a^1)_i = 1$, and carrying out the same procedure as the one to get a^3 from a', we get a'' satisfying $(a'')_i = e_i \; \forall i \in \{1,\ldots,32\} \setminus \{13,\ldots,16\}$.

□

Lemma 6. $\forall a'' \in \{0,1\}^{64} \setminus \{0,d\} : a''_i = e_i \; \forall i \in \{1,\ldots,32\} \setminus \{13,\ldots,16\}, \exists z \in \{0,1\}^{32} : a'' \sim (e,z)$.

Proof. According to Table 2, $[(EP(a)_L)]_4$ corresponds to positions 26, 5, 18, 31, and 2. Since $\{2,5,10,18,26,31\} \cap \{13,\ldots,16\} = \emptyset$, we know $(a''_L)_i = e_i, \; \forall i \in \{2,5,10,18,26,31\}$. Therefore, $[(EP(a)_L)]_4 = (1,1,1,1,1,0) = 62$ and because of Lemma 2 (c), $U([EP((a'')_L)]_j) = \{0,1\}^4$. Thus, considering appropriate $(K^i, K^{i'})$, we get $(e,z) = F^L_{K^1,K^{1'}} \circ \cdots \circ F^L_{K^n,K^{n'}}(a'')$, for some $z \in \{0,1\}^{32}$. □

Corollary 1. $\forall a \in \{0,1\}^{64} \setminus \{0,d\} \; \exists z \in \{0,1\}^{32} : a \sim (e,z)$.

Proof. Considering the chain $a \sim a' \sim a'' \sim (e,z)$, where these elements are as described in the previous lemmata, the result follows. □

Corollary 2. $G_{0,d}$ is transitive on $\{0,1\}^{64} \setminus \{0,d\}$.

Proof. Let $a, a' \in \{0,1\}^{64} \setminus \{0, d\}$, by Lemma 6 and Corollary 1, $\exists z, z' \in \{0,1\}^{32} : a \sim (e, z) \sim (e, z') \sim a'$. □

Corollary 3. G_0 *is transitive on* $\{0,1\}^{64} \setminus \{0\}$.

Proof. Because of Corollary 1, it is enough to show that $\exists g \in G_0$ such that $g(d) \neq d$.

Note that since $g \in G_0$, then $g(d) \neq 0$.

Let $(K, K') \in \mathbb{M} \setminus \mathbb{M}_{d'}$, then $S(K) = S(K')$ and $S(K \oplus d') \neq S(K' \oplus d')$. Therefore, $F^R_{K,K'}(d) = (d_L, d_R \oplus) S(K \oplus d') \oplus S(K' \oplus d') \neq d$, and $F^R_{K,K'} \in G_0$. □

Lemma 7. *If G_0 is transitive on $\{0,1\}^{64} \setminus \{(0, \ldots, 0)\}$ and $G_{0,d}$ is transitive on $\{0,1\}^{64} \setminus \{(0, \ldots, 0), d\}$, then G is 3-transitive on $\{0,1\}^{64}$.*

Proof. It follows immediately from [17] (Theorem 9.1). □

Once we have shown that G is a 3-transitive subgroup of $A_{2^{64}}$, it is not particularly difficult to verify that G is actually equal to the alternating group on 2^{64} points.

Theorem 1. *The round functions of DESL generate the alternating group, i.e., $G = A_{2^{64}}$.*

Proof. We refer to the proof of Theorem 1 in [7], since the same proof applies here. □

4. Applying MRHS to DESL and DES

The previous section focuses on a structural group-theoretic property which does not take the actual number of DESL rounds into account. Subsequently, we studied an algebraic attack against reduced and full round versions of DESL and compared the behavior of the attack with the situation for DES. The underlying question is, to what extent does the modified S-box change the complexity of an algebraic attack?

4.1. Symbol Creation for DESL

Since the structure of DES and DESL is the same, the process for creating the A-parts of MRHS symbols for DESL is the same as that for DES, which is described nicely in [12] (pp. 50–53). The only difference is that the L-part of each symbol will not correspond to a DES S-box, but instead to the DESL S-box. This L-part is given as

$$\begin{bmatrix} 0 & 0 & 0 & 0 & 0 & 0 & 0 & 0 & F & F & F & F & F & F & F & F \\ 0 & 0 & 0 & 0 & F & F & F & F & 0 & 0 & 0 & 0 & F & F & F & F \\ 0 & 0 & F & F & 0 & 0 & F & F & 0 & 0 & F & F & 0 & 0 & F & F \\ 0 & F & 0 & F & 0 & F & 0 & F & 0 & F & 0 & F & 0 & F & 0 & F \\ 3 & 3 & 3 & 3 & 3 & 3 & 3 & 3 & 3 & 3 & 3 & 3 & 3 & 3 & 3 & 3 \\ 5 & 5 & 5 & 5 & 5 & 5 & 5 & 5 & 5 & 5 & 5 & 5 & 5 & 5 & 5 & 5 \\ 8 & 5 & E & 3 & 6 & 9 & 6 & 9 & 6 & 6 & 9 & 9 & A & C & 3 & 5 \\ E & 9 & 4 & 3 & 1 & 6 & F & 8 & 9 & 7 & 2 & C & 6 & C & 9 & 3 \\ 8 & B & D & 6 & 7 & 4 & 8 & 3 & 1 & E & 6 & 1 & C & 9 & 3 & E \\ 6 & 9 & 9 & A & 5 & 9 & 6 & 6 & 6 & 5 & 6 & 9 & 5 & A & A & 9 \end{bmatrix},$$

where each entry is written as standard hex notation to save space. Note that the top six rows correspond to each of the possible inputs to an S-box, and the bottom four rows correspond to the output of the S-box. For example, if the input to the S-box is 000000, then the output is 1110, both being readable from the first column of this matrix. If the input is 000001, then the output is 0101, both being readable from the second column. Further, if the input is 000010, then the output is 0101, and if the input is 000011, the output is 0000.

4.2. Results

For serious ciphers, very often the first MRHS action cycle of agreeing, gluing, and equation extracting (that is, until a guess is called for) will not be sufficient to discover the key, so guesses of the key variables must be committed. Naturally, the fewer guesses required, the better an attack is deemed to be. We give the name δ to the number of key bits we must guess before we discover the whole key through an MRHS attack.

For our attacks, we use a machine called Blue with the following specifications: two quad-core Xeon E5520 2.26 GHz processors (though only one core was used), 24 GB of RAM, using Windows 7 Server (Standard Edition). The ciphertext was 0123456789ABCDEF, and the key was the first 56 bits of the SHA-1 hash of "Katalina" (without quotes).

Under these conditions, DESL was attacked on Blue, varying both the number of rounds of the cipher and the threshold of MRHS. The results are summarized in Table 3, with the note that the threshold listed is actually the base 2 logarithm of the actual threshold, so we always choose a power of 2 for the number of columns each L-part is allowed to grow to.

Table 3. DESL δ on Blue, for varying rounds and thresholds.

Threshold	Rounds of DESL						
	4	6	8	10	12	14	16
20	0	34	36	36	40	38	40
21	0	34	39	37	39	39	42
22	0	33	39	37	38	43	38
23	0	33	38	45	46	48	46

We can see from this data that four rounds of DESL could be handled in the initial turn of an MRHS attack, but things became more complicated with more rounds. For more than six rounds it was not at all guaranteed that an increased threshold would actually help with the computation. Only for twelve rounds did we see an improvement with increased threshold, but once we moved to a threshold of 23, δ increased dramatically.

By way of contrast, DES was attacked on Blue varying the number of rounds and threshold. The results are summarized in Table 4.

Table 4. DES δ on Blue, varying rounds and thresholds.

Threshold	Rounds of DES						
	4	6	8	10	12	14	16
20	1 (+1)	35 (+1)	36 (+0)	36 (+0)	41 (+1)	41 (+3)	40 (+0)
21	0 (+0)	35 (+1)	39 (+0)	37 (+0)	39 (+0)	40 (+1)	39 (−3)
22	0 (+0)	32 (−1)	39 (+0)	37 (+0)	38 (+0)	40 (−3)	38 (+0)
23	0 (+0)	33 (+0)	39 (+1)	43 (−2)	46 (+0)	48 (+0)	46 (+0)

Overall, DESL was about as secure as DES from an MRHS perspective, though there were two occasions where DESL required three more bits to guess before recovering the entire key.

We remark in passing that it was conjectured by Schoonen in [12] (Hypothesis 5.1) that for 7–16 rounds of DES, δ would always be 56 minus the (base 2 logarithm of the) threshold, but Table 4 makes it plain that this was not the case.

5. Conclusions

Unlike DES, the DES Lightweight extension (DESL) uses a single S-box. The security of DESL against a number of common types of attacks has already been argued in the literature. In this work we establish that the round functions of DESL generate the same permutation group as the round functions of DES, namely, the alternating group on 2^{64} points. Moreover, based on our work, DESL appeared to offer comparable resistance to MRHS-based algebraic attacks as DES. Therefore, from these algebraic points of view, DESL has no disadvantage compared to DES, and the structural properties of DESL remain an interesting cryptanalytic topic of study.

Author Contributions: Individual contributions to this article: conceptualization, K.M, R.S., and A.S.C.; methodology, K.M., R.S., and A.S.C.; validation, K.M., R.S., and A.S.C.; formal analysis, A.S.C.; software, K.M. and R.S.; investigation, K.M., R.S., and A.S.C.; resources, R.S. and A.S.C.; writing—original draft preparation, K.M., R.S., and A.S.C.; writing—review and editing, K.M., R.S., and A.S.C.; project administration, R.S. and A.S.C.; funding acquisition, R.S. and A.S.C.

Funding: This research was funded in part by the NATO Science for Peace and Security Programme under grant G5448 and through research project MTM2017-83506-C2-2-P by the Spanish MICINN.

Conflicts of Interest: The authors declare no conflicts of interest. The funders had no role in the design of the study; in the collection, analyses, or interpretation of data; in the writing of the manuscript; or in the decision to publish the results.

References

1. Daley, W.M.; Kammer, R.G. Data Encryption Standard (DES). In *Federal Information Processing Standards Publication*; National Institute of Standards and Technology: Gaithersburg, MD, USA 1999.
2. Leander, G.; Paar, C.; Poschmann, A.; Schramm, K. New Lightweight DES Variants. In *Fast Software Encryption, 14th International Workshop, FSE 2007*; Lecture Notes in Computer Science; Biryukov, A., Ed.; International Association for Cryptologic Research, Springer: New York, NY, USA, 2007; Volume 4593, pp. 196–210.
3. Priyanka, A.A.; Saibal, K.P. A Survey of Cryptanalytic Attacks on Lightweight Block Ciphers. *Int. J. Comput. Sci. Inf. Technol. Secur.* **2012**, *2*, 472–481.
4. Sun, S.; Hu, L.; Qiao, K.; Ma, X.; Shan, J.; Song, L. Improvement on the Method for Automatic Differential Analysis and Its Application to Two Lightweight Block Ciphers DESL and LBlock-s. In Proceedings of the 2015 10th International Workshop on Security Advances in Information and Computer Security, IWSEC, Nara, Japan, 26–28 August 2015; pp. 97–111.
5. Hatzivasilis, G.; Fysarakis, K.; Papaefstathiou, I.; Manifavas, C. A review of lightweight block ciphers. *J. Cryptogr. Eng.* **2018**, *8*, 141–184. [CrossRef]
6. Ji, F.; Zhang, W.; Ding, T. Improving Matsui's Search Algorithm for the Best Differential/Linear Trails and its Applications for DES, DESL and GIFT. Cryptology ePrint Archive, Report 2019/1190. 2019. Available online: http://eprint.iacr.org/2019/1190 (accessed on 14 November 2019).
7. Wernsdorf, R. The One-Round Functions of the DES Generate the Alternating Group. In *Advances in Cryptology—EUROCRYPT '92*; Lecture Notes in Computer Science; Rueppel, R.A., Ed.; Springer: New York, NY, USA, 1993; Volume 658, pp. 99–112.
8. Raddum, H.; Semaev, I. Solving Multiple Right Hand Sides linear equations. *Des. Codes Cryptogr.* **2008**, *49*, 147–160. [CrossRef]
9. Geiselmann, W.; Matheis, K.; Steinwandt, R. PET SNAKE: A Special Purpose Architecture to Implement an Algebraic Attack in Hardware. In *Transactions on Computational Science X*; Lecture Notes in Computer Science; Springer: New York, NY, USA, 2010; Volume 6340, pp. 298–328.
10. Håvard, R.; Zajac, P. MRHS solver based on linear algebra and exhaustive search. *J. Math. Cryptol.* **2018**, *12*, 143–157.
11. Zajac, P. Upper bounds on the complexity of algebraic cryptanalysis of ciphers with a low multiplicative complexity. *Des. Codes Cryptogr.* **2017**, *82*, 43–56. [CrossRef]
12. Schoonen, A.C.C. Multiple Right-Hand Side Equations. Master's Thesis, Department of Mathematics and Computer Science, Eindhoven University of Technology, Eindhoven, The Netherlands, 2008. Available online: http://alexandria.tue.nl/extra1/afstversl/wsk-i/schoonen2008.pdf (accessed on 14 November 2019).
13. Davio, M.; Desmedt, Y.; Fosséprez, M.; Govaerts, R.; Hulsbosch, J.; Neutjens, P.; Piret, P.; Quisquater, J.J.; Vandewalle, J.; Wouters, P. Analytical Characteristics of the DES. In *Advanves in Cryptology—CRYPTO '83*; Chaum, D., Ed.; Plenum Press: New York, NY, USA, 1984; pp. 171–202.

14. Raddum, H. MRHS Equation Systems. In *Selected Areas in Cryptography—SAC 2007*; Lecture Notes in Computer Science; Carlisle Adams, A.M., Wiener, M., Eds.; Springer: New York, NY, USA, 2007; Volume 4876, pp. 232–245.
15. Even, S.; Goldreich, O. DES-Like Functions Can Generate the Alternating Group. *IEEE Trans. Inf. Theory* **1983**, *29*, 863–865. [CrossRef]
16. Foundation, P.S. Python Programming Language—Offical Website. 2010. Available online: http://www.python.org (accessed on 14 November 2019).
17. Wielandt, H. *Finite Permutation Groups*; Academic Press: Cambridge, MA, USA, 1964.

© 2019 by the authors. Licensee MDPI, Basel, Switzerland. This article is an open access article distributed under the terms and conditions of the Creative Commons Attribution (CC BY) license (http://creativecommons.org/licenses/by/4.0/).

Article

Randomness Analysis for the Generalized Self-Shrinking Sequences

Sara D. Cardell [1], Verónica Requena [2], Amparo Fúster-Sabater [3,*] and Amalia B. Orúe [3]

1. Instituto de Matemática, Estatística e Computação Científica, UNICAMP, 13083-859 Campinas-SP, Brazil; scardell@unicamp.br
2. Departamento de Matemáticas, Universidad de Alicante, 03690 Alicante, Spain; vrequena@ua.es
3. Instituto de Tecnologías Físicas y de la Información, CSIC, 28006 Madrid, Spain; amalia.orue@iec.csic.es
* Correspondence: amparo@iec.csic.es

Received: 30 September 2019; Accepted: 23 November 2019; 28 November 2019

Abstract: In cryptography, the property of randomness in pseudo-random generators is very important to avoid any pattern in output sequences, to provide security against attacks, privacy and anonymity. In this article, the randomness of the family of sequences obtained from the generalized self-shrinking generator is analyzed. Moreover, the characteristics, generalities and relationship between the t-modified self-shrinking generator and the generalized self-shrinking generator are presented. We find that the t-modified self-shrunken sequences can be generated from a generalized self-shrinking generator. Then, an in-depth analysis of randomness focused on the generalized sequences by means of complete and powerful batteries of statistical tests and graphical tools is done, providing a useful vision of the behaviour of these sequences and proving that they are suitable to be used in cryptography.

Keywords: generalized self-shrinking generator; t-modified self-shrinking generator; pseudo-random number generator; statistical randomness tests; cryptography

1. Introduction

In cryptography, randomness plays an important role in multiple and diverse applications. Random numbers are employed to generate cryptographic keys, challenges, nonces, to encrypt messages and at different steps of cryptographic algorithms and protocols [1–4].

A pseudo-random number generator is an algorithm for creating a sequence of numbers that is supposed to be indistinguishable from a uniformly chosen random sequence. The sequence is not really random, since it is completely determined by a small set of initial values, called the seed. However, in cryptography, where the security of many cryptographic schemes lies in the quality of pseudorandom generators, it is necessary that the sequences meet the following requirements—(1) the generated sequence must not be distinguished from a truly random sequence; (2) the sequence must be unpredictable; (3) the sequence period must be very large; (4) the key space must be large enough for a brute or exhaustive force attack to be impossible; (5) the design of the generator should be resistant to the specialized attacks reported in the literature.

There is no mathematical proof that ensures the randomness of a bit sequence; however, there exists a huge number of empirical tests to determine if a sequence is random enough and secure to be used in cryptography [5]. If the sequences of a generator pass the statistical tests, then this could be accepted as a generator of random sequences. Otherwise, if several tests fail, it means that the generator is not good and must be rejected. Choosing the correct number of these tests to determine whether the sequence in question can be considered random is a very difficult task since we cannot assure how many tests are needed for it. We have chosen some of those that are considered the most complete randomness tests, like the FIPS (Federal Information Processing Standard) test 140-2 [6],

Diehard Battery of Tests [7], the NIST-SP-800-22 battery test [8] and other tests from the chaos theory, that were presented in References [9,10].

The Generalized Self-Shrinking Generator (GSSG) [11] is fast, easy to be implemented and generates good cryptographic sequences, so it seems suitable for its use in lightweight cryptography and, in general, in low-cost applications. However, the randomness of these sequences has never been analysed with such a complete battery of tests.

In this article, the randomness of the family of sequences obtained from the generalized self-shrinking generator is analyzed. First, the characteristics and generalities of this family of pseudorandom generators have been considered in detail. Then, an in-depth analysis of randomness focused on the generalized sequences by means of complete batteries of statistical tests was done. Tables, figures and graphical representations illustrate the obtained results.

2. Related Work

One of the most accepted designs of Pseudo-Random Number Generator (PRNG) is based on Linear Feedback Shift Registers (LFSR) because LFSRs' sequences can have good statistical properties and their good efficiency in hardware designs. Linear feedback shift registers have been used as basic component of such PRNG but they all have been successfully cryptanalyzed by means of different attacks such as algebraic and correlation attacks, to name a few. Its main weakness is its linearity, which allows the building of a system of equations that solves the parameters used in its design [12].

To avoid these cryptanalytic attacks, new designs use non-linear operations, such as non-linear filtering and sequence decimation, for example. The shrinking generator and the self-shrinking generators are good examples of how to convert a linearly generated sequence into a non-linear one. To do that, different rules, which decimate the LFSR produced sequence in an irregular way, are used. The Shrinking Generator (SG) was firstly proposed in 1993 by Coppersmith, Krawczyk and Mansour [13] and the Self-Shrinking Generator (SSG) in 1994 by Meier and Staffelbach [14].

In Reference [15] a novel generator based on the generalized self-shrinking stream sequence generator (called F-GSS) was proposed, the sequences generated by the F-GSS were analyzed using the NIST statistical test suite, showing that it has good pseudo-random properties.

The Modified Self-Shrinking Generator (MSSG) was proposed by Kanso in Reference [16]. The study of the randomness of this generator was carried out by the NIST statistical test suite and it was demonstrated that sequences of the MSSG have better randomness properties than those of the SSG. In Reference [17] the authors present a new non-periodic random number generator based on the shrinking generator. The randomness of the sequences of the new generator was analyzed by means of Diehard battery of tests, verifying that this new design performs well in this statistical battery of tests.

Tasheva et al. in Reference [18] proposed a variant of the SSG called the p-ary Generalized Self-Shrinking Generator (pGSSG). The authors have studied its randomness using the NIST statistical test suite, later in Reference [19] the balance property of the previously proposed p-ary Generalized Self-Shrinking Generator was studied and it was shown that the generated sequences could be considered as balanced. Erkek and Tuncer in Reference [20] have implemented the SG and Alternating Step Generator on an FPGA Altera Cyclone IV board. Generated numbers in the real time were tested using the NIST statistical test suite. The results have shown that both generators have good statistical properties. In Reference [21] the authors have studied the randomness of the Self-Shrinking generator by means of the d-Monomial test. They have found that there exist some statistical dependencies on certain randomness properties of the generalized SSG and polynomial used in its design. For this reason, they recommend to take special care when choosing the polynomial for the SSG in order to the generator be cryptographically secure. In Reference [22] the author have analyzed a keystream produced by Generalized Shrinking Multiplexing Generator controlled by Ternary m-sequences (GSMG-3m). For randomness analysis they use the NIST statistical test suite, the spectral test and, approximate entropy test. The authors have presented some cryptanalytic work of the proposed generator that prove that GSMG-3m is more secure than the Shrinking Generator.

As can be seen, there are few works that have deeply studied the randomness of the sequences generated by the different families of shrinking generators through several statistical test batteries such as those presented in this paper.

3. Preliminaries

In order to the work be self-contained, some basic concepts concerning binary sequences as well as sequence generators based on irregular decimation are introduced. All of them will be used throughout the paper.

As has been said previously, the security of many cryptographic algorithms is based on a well designed random and pseudorandom generators. It is worth mentioning that the design of reliable and secure pseudorandom number generators is an open problem and an intensive field of research in cryptography nowadays [23–28]. The family of shrinking generators is one of the most analyzed PRNG in the literature due to its performance and security when it is well designed [4,21,22,29–31].

3.1. PN-Sequences

Let $\mathbb{F}_2 = \{0, 1\}$ be the Galois field. Consider $\{a_i\}_{i \geq 0} = \{a_0, a_1, a_2 \ldots\}$ a binary sequence with $a_i \in \mathbb{F}_2$, for $i = 0, 1, 2, \ldots$ We say the sequence $\{a_i\}_{i \geq 0}$ is periodic if there exists an integer T, called period, such that $a_{i+T} = a_i$, for all $i \geq 0$. In the sequel, all the sequences considered will be binary sequences and the symbol $+$ will denote the Exclusive-OR (XOR) logic operation.

Let r be a positive integer and let $d_1, d_2, d_3, \ldots, d_r$ be constant coefficients with $d_j \in \mathbb{F}_2$. A binary sequence $\{a_i\}_{i \geq 0}$ satisfying the relation:

$$a_{i+r} = d_r a_i + d_{r-1} a_{i+1} + \cdots + d_3 a_{i+r-3} + d_2 a_{i+r-2} + d_1 a_{i+r-1}, \quad i \geq 0, \qquad (1)$$

is called a (r-th order) linear recurring sequence (LRS) in $\mathbb{F}_2$. The terms $\{a_0, a_1, \ldots, a_{r-1}\}$ are referred to as the initial terms and define the construction of the sequence uniquely.

The monic polynomial:

$$p(x) = d_r + d_{r-1}x + \cdots + d_3 x^{r-3} + d_2 x^{r-2} + d_1 x^{r-1} + x^r \in \mathbb{F}_2[x]$$

is called the characteristic polynomial of the linear recurring sequence and $\{a_i\}_{i \geq 0}$ is said to be generated by $p(x)$.

Linear recurring sequences can be generated using Linear Feedback Shift Registers (LFSRs) [5,12,32]. In fact, an LFSR can be defined as an electronic device with r memory cells (stages) with binary content. At every clock pulse, the binary element of each stage is shifted to the adjacent stage as well as a new element is computed through the linear feedback to fill the empty stage (see Figure 1). The LFSR has maximal-length if the characteristic polynomial of the LFSR is primitive. Its output sequence is called PN-sequence (Pseudo-Noise sequence) and has period $T = 2^r - 1$, see Reference [32].

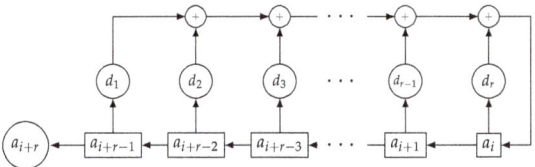

Figure 1. LFSR of length r.

The linear complexity, *LC*, of a sequence $\{a_i\}_{i\geq 0}$ is defined as the length of the shortest LFSR that generates such a sequence or, equivalently, as the lowest order linear recurrence relationship that generates such a sequence.

In cryptographic terms, the linear complexity must be as large as possible as *LC* defines the minimum piece of the sequence needed to get the whole sequence.

A simple result that will be useful in the next section is introduced below.

Lemma 1. *Let $\{a_i\}_{i\geq 0}$ be a PN-sequence with period T. Then, the sequence $\{u_i\}$ such that $u_i = \sum_{k=0}^{t-2} a_{t \cdot i + k}$ is again a PN-sequence with the same period T iff $\gcd(T,t) = 1$.*

Proof. The sequence $\{a_{t \cdot i}\}$ is a PN-sequence iff $\gcd(T,t) = 1$, see Reference [32] (pag. 78). The sequences $\{a_{t \cdot i + k}\}$ for $k = 0, \ldots, t-2$ are shifted versions of $\{a_{t \cdot i}\}$ with different starting points. If we XOR a PN-sequence with a shifted sequence of itself, then we have the same PN-sequence but starting at a different bit [32] (Theorem 4.3–4.5). Thus, $\{u_i\}$ is the same sequence as $\{a_{t \cdot i}\}$ except for the starting point, that is, $\{u_i\} = \left\{\sum_{k=0}^{t-2} a_{t \cdot i + k}\right\} = \{a_{t \cdot i + D}\}$ where $D < T$ is a positive integer. □

3.2. Modified Self-Shrinking Generator (MSSG)

Decimation is a very habitual technique to produce pseudo-random sequences with cryptographic applications [33,34]. In practice, the underlying idea in this kind of generators is the irregular decimation of a PN-sequence according to the bits of another.

The Modified Self-Shrinking Generator (MSSG) introduced by Kanso in Reference [16] is a modification of the well-known Self-Shrinking Generator (SSG) [14]. Indeed, in the MSSG the PN-sequence $\{a_i\}_{i\geq 0}$ generated by a maximal-length LFSR is self-decimated. The decimation rule is very simple and can be described as follows: given three consecutive bits $\{a_{3i}, a_{3i+1}, a_{3i+2}\}$, $i = 0, 1, 2, \ldots$, the output sequence $\{s_j\}_{j\geq 0}$ is computed as

$$\begin{cases} \text{If } a_{3i} + a_{3i+1} = 1 \text{ then } s_j = a_{3i+2}, \\ \text{If } a_{3i} + a_{3i+1} = 0 \text{ then } a_{3i+2} \text{ is discarded.} \end{cases}$$

The output sequence $\{s_j\}_{j\geq 0}$ is known as the Modified Self-Shrunken sequence (MSS-sequence). If *L* is the length of the maximal-length LFSR that generates $\{a_i\}_{i\geq 0}$, then the linear complexity *LC* of the corresponding MSS-sequence satisfies:

$$2^{\lfloor \frac{L}{3} \rfloor - 1} \leq LC \leq 2^{L-1} - (L-2),$$

and the period *T* of the sequence, when *L* is odd, satisfies:

$$2^{\lfloor \frac{L}{3} \rfloor} \leq T \leq 2^{L-1},$$

as proved in Reference [16]. As usual, the key of this generator is the initial state of the LFSR that generates $\{a_i\}_{i\geq 0}$. The characteristic polynomial of such a register is also recommended to be part of the key.

Example 1. *Consider the LFSR of length $L = 4$ with characteristic polynomial $q(x) = x^4 + x + 1$ and the initial state $\{1\ 1\ 1\ 1\}$. The corresponding PN-sequence is given by $\{111100010011010\ldots\}$ with period $T = 2^4 - 1$.*

The MSS-sequence is obtained as follows:

R: 11 ⟨10 **0**⟩ 01 **0** 01 **1** 01 **0** 11 ⟨10 **0**⟩ ...
+:   0   1   1   1   1   0   1

The obtained sequence $\{s_j\} = \{0\,0\,1\,0\,\ldots\}$ (encircled bits) has period $T = 4$ and it can be checked that its characteristic polynomial is $p_4(x) = 1 + x^4$. Thus, the linear complexity of this MSS-sequence is $LC = 4$.

In Reference [30], the authors showed that the sequences produced by this generator are contained in the family of sequences generated by the generalized self-shrinking generator.

3.3. The Generalized Self-Shrinking Generator (GSSG)

In this subsection, we introduce the most representative generator in this family of decimation-based sequence generators, that is, the Generalized Self-Shrinking Generator (GSSG) [11]. In fact, the sequences produced by this generator include the sequences produced by the generators previously described.

Let $\{a_i\}_{i\geq 0}$ be an PN-sequence produced by a maximal-length LFSR with L stages. Let $G = [g_0, g_1, g_2, \ldots, g_{L-1}] \in \mathbb{F}_2^L$ be an L-dimensional binary vector and $\{v_i\}_{i\geq 0}$ a sequence defined as: $v_i = g_0 a_i + g_1 a_{i-1} + g_2 a_{i-2} + \cdots + g_{L-1} a_{i-L+1}$. For $i \geq 0$, the decimation rule is defined as follows:

$$\begin{cases} \text{If } a_i = 1 \text{ then } s_j = v_i, \\ \text{If } a_i = 0 \text{ then } v_i \text{ is discarded.} \end{cases}$$

The output sequence generated $\{s_j\}_{j\geq 0}$ associated with G, denoted by $s(G)$, is called the Generalized Self-Shrunken sequence (GSS-sequence).

When G ranges over $\mathbb{F}_2^L$, then $\{v_i\}$ corresponds to the $2^L - 1$ possible shifts of $\{a_i\}$, that is, the sequence $\{v_i\}$ is a shifted version of the PN-sequence $\{a_i\}$. Moreover, we obtain the family of generalized self-shrunken sequences based on the PN-sequence $\{a_i\}_{i\geq 0}$ given by the set of sequences denoted by $S(a) = \{s(G) | G \in \mathbb{F}_2^L\}$. In Table 1, the algorithm to compute these sequences is shown (Algorithm 1).

Table 1. Algorithm to compute the GSS-sequences.

Algorithm 1: Constructing the family of GSS-sequences
Input: Primitive polynomial $p(x)$ and initial state a
01: Compute the PN-sequence $\{a_i\}$.
02: Set $T = 2^L - 1$ the period of the PN-sequence
03: **for** $p = 1$ to T **do**
04: Set $\{v_i\}$ the shifted version of $\{a_i\}$ by p positions
06: **for** $k = 0$ to $T - 1$ **do**
06 Initialize sequence $\{s_j^p\}$
07: **if** $a_k = 1$ **do**
08: Add v_k as new bit of the sequence $\{s_j^p\}$
09: **endif**
10: **end for**
11: **end for**
Output: $\{s_j^p\}$ GSS-sequences, $p = 1, \ldots, T$.

Example 2. *Consider the primitive polynomial $p(x) = 1 + x^3 + x^4$ and the corresponding PN-sequence $\{a_i\}_{i\geq 0} = \{111101011001000\}$ We can construct the GSS sequences shown in Table 2. The underlined bits in the different sequences $\{v_i\}_{i\geq 0}$ are the digits of the corresponding $\{s(G)\}$ sequences. The PN-sequence $\{a_i\}_{i\geq 0}$ is written at the bottom of the table.*

Table 2. Family of Generalized Self-Shrunken sequences generated by $p(x) = 1 + x^3 + x^4$.

G	$\mathcal{G}$	$\{v_i\}$ Sequence	Generalized Sequence
0	0000	000000000000000	00000000
1	0001	000111101011001	00011011
2	0010	001111010110010	00111100
3	0011	001000111101011	00100111
4	0100	011110101100100	01110010
5	0101	011001000111101	01101001
6	0110	010001111010110	01001110
7	0111	010110010001111	01010101
8	1000	111101011001000	11111111
9	1001	111010110010001	11100100
10	1010	110010001111010	11000011
11	1011	110101100100011	11011000
12	1100	100011110101100	10001101
13	1101	100100011110101	10010110
14	1110	101100100011110	10110001
15	1111	101011001000111	10101010
		111101011001000	

4. The *t*-Modified Self-Shrinking Generator

A generalization of GSSG, the *t*-Modified Self-Shrinking Generator (*t*-MSSG) was introduced by Cardell et al. in Reference [31] and can be described as follows. Consider a maximal-length LFSR with L stages that generates the PN-sequence $\{a_i\}_{i \geq 0}$. The *t*-modified self-shrinking generator, with $(t = 2, 3, \ldots, 2^L - 2)$, can be constructed making use of a very simple decimation rule.

Given t consecutive bits $\{a_{t \cdot i}, a_{t \cdot i+1}, a_{t \cdot i+2}, \ldots, a_{t \cdot i+(t-1)}\}$ of the PN-sequence, the output sequence of this generator $\{s_j\}_{j \geq 0}$ is known as the *t*-Modified Self-Shrunken sequence (*t*-MSS-sequence) and computed as follows:

$$\begin{cases} \text{If } \sum_{j=0}^{t-2} a_{t \cdot i+j} = 1 \text{ then } s_j = a_{t \cdot i+(t-1)}, \\ \text{If } \sum_{j=0}^{t-2} a_{t \cdot i+j} = 0 \text{ then } a_{t \cdot i+(t-1)} \text{ is discarded.} \end{cases} \quad (2)$$

Notice that the value $t = 2$ gives rise to the self-shrinking generator [14] while the value $t = 3$ defines the modified self-shrinking generator. In Table 3 the algorithm to compute this sequence is presented (Algorithm 2). Characteristics and generalities of the *t*-MSS-sequences can be found in Reference [31].

Table 3. Algorithm to compute the *t*-MSS-sequence.

Algorithm 2: Constructing the *t*-MSS-sequence
Input: Primitive polynomial $p(x)$, initial state $\mathbf{a}$ and t
01: Compute the PN-sequence $\{a_i\}$.
02: Set $T = 2^L - 1$ the period of the PN-sequence
03: **for** $k = 0$ **to** $T - 1$ **do**
04:     Initialize sequence $\{s_j\}$
05:     **if** $\sum_{j=0}^{t-2} a_{t \cdot k+j} = 1$ **do**
06:         Add $a_{t \cdot k+(t-1)}$ as new bit of the sequence $\{s_j\}$
07:     **endif**
08: **end for**
Output: $\{s_j\}$ *t*-MSS-sequence.

Relationship between t-Modified Self-Shrunken Sequences and Generalized Self-Shrunken Sequences (GSS-Sequences)

Now, we analyse the close relationship between *t*-Modified Self-Shrunken sequences (*t*-MSS-sequences) and Generalized Self-Shrunken sequences (GSS-sequences).

In Theorem 1 of Reference [30], they analyse the relationship between modified self-shrunken sequences and generalized self-shrunken sequences with a result similar to the following:

Theorem 1. *The t-MSS-sequence as a result of self-decimating a PN-sequence with characteristic polynomial $q(x)$ of degree L and $\gcd(T,t) = 1$, can be generated from a generalized self-shrinking generator with a primitive polynomial $p(x)$ of the same degree L.*

Proof. Let $\{a_i\}$ be a PN-sequence with characteristic polynomial $q(x)$ of degree L which is self-decimated. In order to generate the *t*-MSS-sequence, sets of *t* bits $\{a_{t \cdot i}, a_{t \cdot i+1}, a_{t \cdot i+2}, \ldots, a_{t \cdot i+(t-1)}\}$, $(i \geq 0)$ have to be taken. Applying the decimation rule defined in 2, if $\sum_{k=0}^{t-2} a_{t \cdot i+k} = 1$, the bit $a_{t \cdot i+(t-1)}$ is kept. Otherwise, it is discarded. According to Lemma 1, the sequence $\{u_i\}$ defined as $u_i = \sum_{k=0}^{t-2} a_{t \cdot i+k} = a_{t \cdot i+D}$ is obtained by decimating the sequence $\{a_i\}$ by distance t.

Since $\gcd(T,t) = 1$, according to Reference [32], we have that $\{u_i\}$ is a PN-sequence generated by a primitive polynomial $p(x)$ of the same degree, L.

Also, if the sequence $\{v_i\}$ is taken, with $v_i = a_{t \cdot i+(t-1)}$, this means that the sequence $\{a_i\}$ is being decimated again by the distance t. As before, we have that $\{v_i\}$ is also a PN-sequence with primitive polynomial $p(x)$ [32].

In order to obtain the *t*-MSS-sequence, the *t*-MSSG decimation rule is applied to the sequences $\{u_i\}$ and $\{v_i\}$. As both sequences are shifted versions of the PN-sequence $\{a_i\}$, we can generate such a *t*-MSS-sequence by a GSSG with characteristic polynomial $p(x)$. □

As a result of the previous theorem, we have that:

Corollary 1. *If $t = 2, 4, \ldots, 2^{L-1}$, then the t-MSS-sequence is generated as a generalized sequence with the same primitive polynomial $q(x)$.*

Proof. It follows from the following idea: the sequence $\{a_{t \cdot i}\}$ is a shifted version of the PN-sequence $\{a_i\}$ when $t = 1, 2, \ldots, 2^{L-1}$, see Reference [32] (pag. 76). □

The next theorem gives us the primitive polynomial $p(x)$ that we need in Theorem 1 in order to the GSSG generates the *t*-MSS-sequence obtained with a characteristic polynomial $q(x)$.

Theorem 2. *When $\gcd(T,t) = 1$, the primitive polynomial $p(x)$ in Theorem 1 is:*

$$p(x) = (x + \alpha^t)(x + \alpha^{2t})(x + \alpha^{4t}) \cdots (x + \alpha^{t \cdot 2^{L-1}}),$$

where $\alpha \in \mathbb{F}_{2^L}$ is a root of $q(x)$.

Proof. The primitive polynomial $q(x)$ can be expressed as:

$$q(x) = (x + \alpha)(x + \alpha^2)(x + \alpha^4) \cdots (x + \alpha^{2^{L-1}}),$$

where $\alpha \in \mathbb{F}_{2^L}$ is a primitive element in such a field as well as a root of $q(x)$. Furthermore, any element of the PN-sequence $\{a_i\}$ is obtained as:

$$a_i = A_0 \alpha^i + A_0^2 \alpha^{2i} + A_0^4 \alpha^{4i} + \cdots + A_0^{2^{L-1}} \alpha^{2^{L-1} i},$$

with $A_0 \in \mathbb{F}_{2^L}$ [35]. When $A_0 = 1$, it is said that the PN-sequence is in its characteristic phase.

The following sequence is obtained:

$$\{a_0, a_t, a_{2t}, \ldots, a_{t \cdot 2^{L-1}}, \ldots\},$$

decimating the sequence $\{a_i\}$ by distance t. That is a PN-sequence (since $\gcd(T,t) = 1$) and each one of its bits can be computed as:

$$a_{t \cdot i} = A_0 \alpha^{t \cdot i} + A_0^2 \alpha^{2t \cdot i} + A_0^4 \alpha^{4t \cdot i} + \cdots + A_0^{2^{L-1}} \alpha^{t \cdot 2^{L-1} i}.$$

If $u_i = a_{t \cdot i}$ and $\beta = \alpha^t$, then any element of the PN-sequence $\{u_i\}$ can be computed as follows:

$$u_i = A_0 \beta^i + A_0^2 \beta^{2i} + A_0^4 \beta^{4i} + \cdots + A_0^{2^{L_2}-1} \beta^{2^{L_2-1} i}.$$

Therefore, the characteristic polynomial of the PN-sequence $\{u_i\}$ is,

$$p(x) = (x + \beta)(x + \beta^2)(x + \beta^4) \cdots (x + \beta^{2^{L-1}}),$$

or, equivalently,

$$p(x) = (x + \alpha^t)(x + \alpha^{2t})(x + \alpha^{4t}) \cdots (x + \alpha^{t \cdot 2^{L-1}}).$$

□

Lemma 2. *Given a PN-sequence $\{a_i\}$ of prime period $T = 2^L - 1$ and characteristic polynomial $q(x)$ of degree L, then sequence $\{a_{t \cdot i}\}$ is a PN-sequence of period T, for any t.*

Proof. According to Reference [32], $\{a_{t \cdot i}\}$ is a PN-sequence of period T if $\gcd(T,t) = 1$. Since T is prime, then $\gcd(T,t) = 1$ for any t. □

Theorem 3. *Given a PN-sequence with period prime $T = 2^L - 1$ and $q(x)$ characteristic polynomial of degree L, then the t-MSS-sequence obtained for any t is a generalized sequence generated with a primitive polynomial of degree L.*

Proof. The proof follows the same reasoning used in Theorem 1 and Lemma 2. □

Example 3. *Given $p(x) = 1 + x^2 + x^5$, the period of the PN-sequence $\{a_i\}$ is $T = 31$, which is a prime number. Table 4 shows all the t-MSS-sequences generated with this polynomial. All of them are generalized sequences obtained from a primitive polynomial $q(x)$ of degree 5. It is important to mention that some generalized sequences can be generated using different primitive polynomials. For example, the generalized sequence $\{101010101010101\}$ can be obtained using any primitive polynomial of degree 5.*

Table 4. t-MSS-sequences obtained with $q(x) = 1 + x^2 + x^5$.

t	t-MSS-Sequence	LC	$p(x)$
2	1101100110100001	13	$1 + x^2 + x^5$
3	1100100101110010	12	$1 + x^2 + x^3 + x^4 + x^5$
4	1000111001011100	13	$1 + x^2 + x^5$
5	1000111011000101	13	$1 + x + x^2 + x^4 + x^5$
6	0100111011011000	13	$1 + x^2 + x^3 + x^4 + x^5$
7	0001011111001010	12	$1 + x + x^2 + x^3 + x^5$
8	0110101111010000	12	$1 + x^2 + x^5$
9	1111000001011010	10	$1 + x + x^2 + x^4 + x^5$
10	0110001001011110	13	$1 + x + x^2 + x^4 + x^5$
11	0011010010110011	13	$1 + x + x^3 + x^4 + x^5$
12	1010000101111100	12	$1 + x^2 + x^3 + x^4 + x^5$
13	0010011001001111	13	$1 + x + x^3 + x^4 + x^5$
14	1001000110111100	13	$1 + x + x^2 + x^3 + x^5$
15	1110010000110110	13	$1 + x^3 + x^5$
16	1101000010100111	12	$1 + x^2 + x^5$
17	0100111110100001	12	$1 + x^2 + x^3 + x^4 + x^5$
18	1111010011001000	13	$1 + x + x^2 + x^4 + x^5$
19	0111101011000001	12	$1 + x + x^2 + x^3 + x^5$
20	1110011000110100	13	$1 + x + x^2 + x^4 + x^5$
21	0101111100001010	10	$1 + x + x^3 + x^4 + x^5$
22	1001100001011011	13	$1 + x + x^3 + x^4 + x^5$
23	0001011011011010	11	$1 + x^3 + x^5$
24	0110011110100100	13	$1 + x^2 + x^3 + x^4 + x^5$
25	0011011011100100	13	$1 + x + x^2 + x^3 + x^5$
26	1100011001110010	13	$1 + x + x^3 + x^4 + x^5$
27	0010111100011100	11	$1 + x^3 + x^5$
28	0111000100111010	13	$1 + x + x^2 + x^3 + x^5$
29	1010000111000111	11	$1 + x^3 + x^5$
30	1010101010101010	2	$1 + x^3 + x^5$

Next, the relationship between t-MSS-sequences and GSS-sequences is analyzed from other point of view, using the cyclotomic cosets given in Reference [32].

Next, we introduce the concept of cyclotomic coset mod $(2^L - 1)$ [32] and some of its properties:

Definition 1 (Cyclotomic cosets mod $(2^L - 1)$). : Let $\mathbb{Z}_{2^L} = \{0, 1, 2, \ldots, 2^L - 1\}$. We define the equivalence relation R between $t_1, t_2 \in \mathbb{Z}_{2^L}$ as follows: t_1 R t_2 if there exists an integer j, $0 \leq j \leq L - 1$, such that

$$2^j \cdot t_1 = t_2 \bmod (2^L - 1).$$

$\mathbb{Z}_{2^L}^*$ is partitioned into resultant equivalence classes called the cyclotomic cosets mod $(2^L - 1)$.

The smallest integer i in any equivalence class is defined as the leader of the coset and is denoted by C_i. The cardinal of a coset is L or a proper divisor of L. The characteristic polynomial of a cyclotomic coset C_i is the polynomial $P_{C_i}(x) = (x + \alpha^i)(x + \alpha^{2i})\ldots(x + \alpha^{2^{r-1}i})$, where the degree r ($r \leq L$) equals the cardinal of the coset C_i and α is a root of the LFSR characteristic polynomial.

Following [32] (Chapter 4), C_i is a proper coset if $\gcd(2^L - 1, i) = 1$, therefore in this case, $P_{C_i(x)}$ is a primitive polynomial, which is a remarkable property because if $P_{C_i(x)}$ is a primitive polynomial the sequence generated by the basic LFSR is as large as possible.

Example 4. *Consider the set $\mathbb{Z}_{2^5}^*$. Notice that $2^5 - 1$ is a primer integer. There are six cyclotomic cosets given by:*

$$C_1 = \{1, 2, 4, 8, 16\} \qquad C_5 = \{5, 10, 20, 9, 18\} \qquad C_{11} = \{11, 22, 13, 26, 21\}$$
$$C_3 = \{3, 6, 12, 24, 17\} \qquad C_7 = \{7, 14, 28, 25, 19\} \qquad C_{15} = \{15, 30, 29, 27, 23\}$$

In this case, all cosets are proper cosets and have cardinal 5. If $q(x) = 1 + x^2 + x^5$ is considered the characteristic polynomial of the LFSR, then the corresponding characteristic polynomial of the cosets are given in Table 5. Since all cosets are proper, all the characteristic polynomials are primitive of degree 5.

Table 5. Characteristic polynomial of cyclotomic cosets.

C_i	$P_{C_i}(x)$
C_1	$1 + x^2 + x^5$
C_3	$1 + x^2 + x^3 + x^4 + x^5$
C_5	$1 + x + x^2 + x^4 + x^5$
C_7	$1 + x + x^2 + x^3 + x^5$
C_{11}	$1 + x + x^3 + x^4 + x^5$
C_{15}	$1 + x^3 + x^5$

Theorem 4. *Consider a PN-sequence of period prime $T = 2^L - 1$ and its characteristic polynomial $q(x)$ of degree L, then both t-MSS-sequences obtained for any t_1 and t_2 are generalized sequences produced by the same polynomial of degree L iff t_1 and t_2 belong to the same coset.*

Proof. According to the proof of Theorem 1, a *t*-MSS-sequence is obtained decimating the sequence $\{a_{t \cdot i}\}$ with a shifted version of itself, that is, as a generalized sequence. According to [32] (Theorem 5.5), $\{a_{t_1 \cdot i}\}$ and $\{a_{t_2 \cdot i}\}$ are shifted versions of the same PN-sequence iff t_1 and t_2 belong to the same coset. Thus, the decimation rule is applied to two shifted versions of the same PN-sequence and, consequently, a generalized sequence has been generated. □

As already mentioned, Table 4 shows all the *t*-MSS-sequences generated by $q(x) = 1 + x^2 + x^5$ and $t = 2, 3, \ldots, 30$. Notice that when t_1 and t_2 are in the same coset, then the corresponding *t*-MSS-sequences are generalized GSS-sequences produced by the same polynomial (characteristic polynomial of the LFSR).

Furthermore, reciprocal polynomials generate sometimes the same sequences with different starting points. For example, the generalized sequence produced with $t = 29$ can be also generated as a generalized sequence using $q(x) = 1 + x^2 + x^5$.

In the following example (Example 5), notice that when $2^L - 1$ is not prime, different types of cyclotomic cosets can be obtained [31].

Example 5. *Consider the set $\mathbb{Z}_{2^4}^*$. Notice that $2^4 - 1$ is not a prime number. There are 4 cyclotomic cosets given by:*

$$C_1 = \{1, 2, 4, 8\} \qquad C_5 = \{5, 10\}$$
$$C_3 = \{3, 6, 12, 9\} \qquad C_7 = \{7, 14, 13, 11\}$$

In this case, C_1 and C_7 are proper cosets and C_5 and C_3 are improper cosets. Therefore, we know that the $P_{C_1}(x)$ and $P_{C_7}(x)$ are primitive polynomials. Consider $q(x) = 1 + x + x^4$ as the characteristic polynomial of the LFSR. Then, the characteristic polynomial of the cosets are given in Table 6. We can check that $P_{C_1}(x)$ and $P_{C_7}(x)$ are primitive polynomials of degree 4 and $P_{C_3}(x)$ is an irreducible polynomial of degree 4. The polynomial $P_{C_5}(x)$ is a primitive polynomial of degree 2.

Table 6. Characteristic polynomial of cyclotomic cosets.

C_i	$P_{C_i}(x)$
C_1	$1 + x + x^4$
C_3	$1 + x + x^2 + x^3 + x^4$
C_5	$1 + x + x^2$
C_7	$1 + x^3 + x^4$

Theorem 5. *Given a PN-sequence of period $T = 2^L - 1$ and characteristic polynomial $q(x)$ of degree L, then both t-MSS-sequences obtained for any t_1 and t_2 are GSS-sequences generated by the same primitive polynomial of degree L iff t_1 and t_2 belong to the same proper coset.*

Proof. If the coset C_i, such that $t_1, t_2 \in C_i$ is proper, it means that $\gcd(t_1, T) = \gcd(t_2, T) = 1$. The rest follows from previous results. □

Remark 1. *When $\gcd(t, T) \neq 1$, the corresponding t-MSS-sequence is a generalized sequence iff $P_{C_t}(x)$ is a primitive polynomial of degree equal to $|C_i|$ (cardinal of C_i).*

Since, under not very restrictive conditions, the GSS-sequences include the other sequences produced by decimation-based generators, our randomness analysis focuses on this class of binary sequences.

In Table 7, we summarize the three more popular decimation-based sequence generators with the bounds for their periods and their linear complexities that were discussed in this work.

Table 7. Summary of the main characteristics of the three decimation-based generators discussed in this work.

Generator	Decimation Rule	Period	LC
Modified self-shrinking (MSSG), [16]	Given three consecutive bits, the output sequence $\{s_j\}_{j \geq 0}$ is computed as: If $a_{3i} + a_{3i+1} = 1$ then, $s_j = a_{3i+2}$. If $a_{3i} + a_{3i+1} = 0$ then, a_{3i+2} is discarded.	$2^{\lfloor L/3 \rfloor} \leq T \leq 2^{L-1}$	When L odd: $2^{\lfloor L/3 \rfloor - 1} \leq LC \leq 2^{L-1} - (L-2)$.
Generalized self-shrinking (GSSG), [11]	Let $\{a_i\}_{i \geq 0}$ be an PN-sequence generated by a maximal-length LFSR with L stages. Let G be an L-dimensional binary vector $G = [g_0, g_1, g_2, ..., g_{L-1}] \in \mathbb{F}_2^L$ and $\{v_i\}_{i \geq 0}$ a sequence defined as: $v_i = g_0 a_i + g_1 a_{i-1} + g_2 a_{i-2} + \cdots + g_{L-1} a_{i-L+1}$. For $i \geq 0$, the decimation rule is: If $a_i = 1$ then $s_j = v_i$. If $a_i = 0$ then v_i is discarded.	$T = 2^r$, with $r \leq L - 1$	$LC \leq 2^{L-1} - (L-2)$.
t-modified self-shrinking (t-MSSG), [31]	Given t consecutive bits, the output sequence $\{s_j\}_{j \geq 0}$ is computed as: If $\sum_{j=0}^{t-2} a_{t \cdot i + j} = 1$ then, $s_j = a_{t \cdot i + (t-1)}$. If $\sum_{j=0}^{t-2} a_{t \cdot i + j} = 0$ then, $a_{t \cdot i + (t-1)}$ discarded.	If $\gcd\{2^L - 1, t\} = 1$ or P_{C_t} is primitive with degree $\|C_i\| : T = 2^r$, with $r \leq L - 1$. Other cases are not cryptographic relevant.	If $\gcd\{2^L - 1, t\} = 1$ or P_{C_t} is primitive with degree $\|C_i\| :$ $LC \leq 2^{L-1} - (L-2)$. Other cases are not cryptographic relevant.

5. Statistical Randomness Analysis

In this section, an exhaustive analysis of randomness of the proposed GSS-sequences is presented by using different batteries of statistical tests to study their behaviour. Some graphical tools from chaos theory have been used [9,10], for example, return maps, chaos game, Lyapunov exponent, and so forth. The generator and the battery of tests were implemented with Matlab 9.1 (2017) in a Windows 10 environment in a 64 bits PC with CPU Intel Core i7-870, at 2.93 GHz.

For our study, GSS-sequences $s(G)$ are generated from PN-sequences coming from maximal-length LFSRs with characteristic polynomials of degree less than or equal to 27. Every one of these sequences has passed perfectly the Diehard battery of tests, considered one of the most important and powerful tool for randomness study.

Furthermore, the family of GSS-sequences is analysed with the family of statistical tests FIPS 140-2, provided by the National Institute of Standards and Technology (NIST), as well as with the Lempel-Ziv Compression Test. In both cases the sequences have passed the tests.

5.1. Graphical Testing

In this section, the main graphical tests used in Reference [9], are applied to the GSS-sequences, from which their cryptographic properties can be analyzed.

The results obtained for GSS-sequences $s(G)$ of length 2^{23} bits, is presented. These sequences are generated by the GSSG from a maximal-length LFSR with the 24-degree characteristic polynomial $p(x) = x^{24} + x^{20} + x^{17} + x^{13} + x^{10} + x^7 + x^4 + x^2 + 1$ and whose initial state is the identically 1 vector of length 24.

The tests were performed with 2^{23} bit sequences. Most of the tests works associating every eight bits in an octet, obtaining sequences of 2^{20} samples of 8 bits; with the exception of the Linear complexity test that works with just one bit and the Chaos game that works associating the bits two by two.

Next, the results of graphical tests to study the randomness of our sequences is shown.

1. *Return map*

Return map [10] tries to measure visually the entropy of the sequence, that is, allows to detect the existence of some useful information about the parameters used in the design of pseudo-random generators [36]. This test, that customarily is used in theory of dynamic systems, is also a powerful tool in cryptanalysis.

Basically, it consists of a graph of the points of the sequence x_t as a function of x_{t-1} and, under certain conditions, allows us to obtain the value of the parameters of a pseudo-random sequence, defeating the security of the cryptosystem under analysis. The result should be a distribution of points where you cannot guess neither trends, nor figures, nor lines, nor symmetry, nor patterns.

Figure 2 shows the return map of our GSS-sequence as a disordered cloud, which does not provide any useful information for its cryptanalysis.

Figure 3a,b are the return applications of two imperfect generators where the lack of randomness can be neatly observed. Indeed, these maps present clear patterns that permit to determine the generator function and the parameter values.

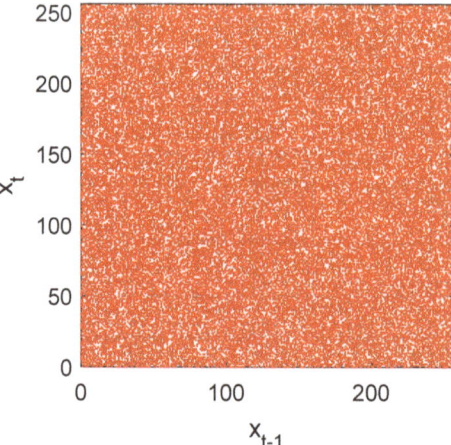

Figure 2. Return map of GSS-sequence of 2^{23} bits. It provides no information about the parameters of the generator.

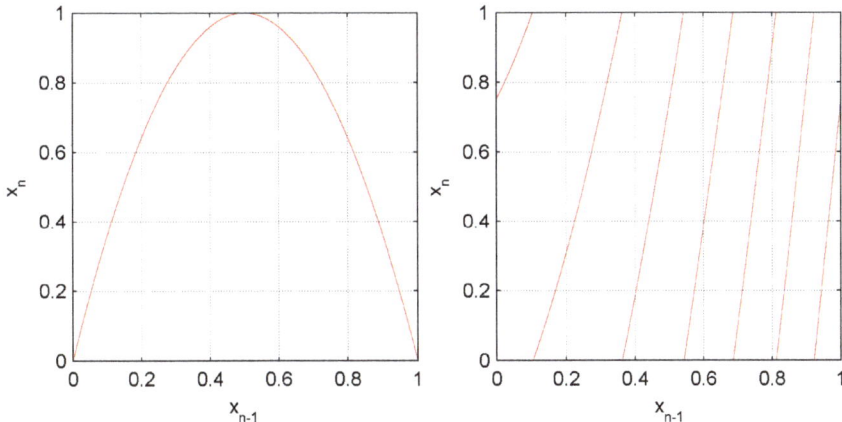

(a) Return map of logistic generator.

(b) Return map of quadratic generator.

Figure 3. Return maps of imperfect generators. The parameter values can be deduced by inspection of the return map.

2. Linear Complexity

The linear complexity (*LC*) is considered as a measure of the unpredictability of a pseudo-random sequence and is a widely used metric of the security of a keystream sequence [37]. We have used the Berlekamp-Massey algorithm [38] to compute this parameter. If the characteristic polynomial of the LFSR is primitive [32], then it is known as maximal-length LFSR; moreover, its output sequence has period $T = 2^L - 1$, where L is the degree of the characteristic polynomial.

LC must be as large as possible, that is, its value has to be very close to half the period [39], $LC \simeq T/2$. From Figure 4a, it can be deduced that the value of the linear complexity of the first 20,000 bits of the sequence is just half its length, 10,000 and, from Figure 4b is observed that *LC* is *irregularly* close to the $\frac{l}{2}$-line, being l the length of the sequence.

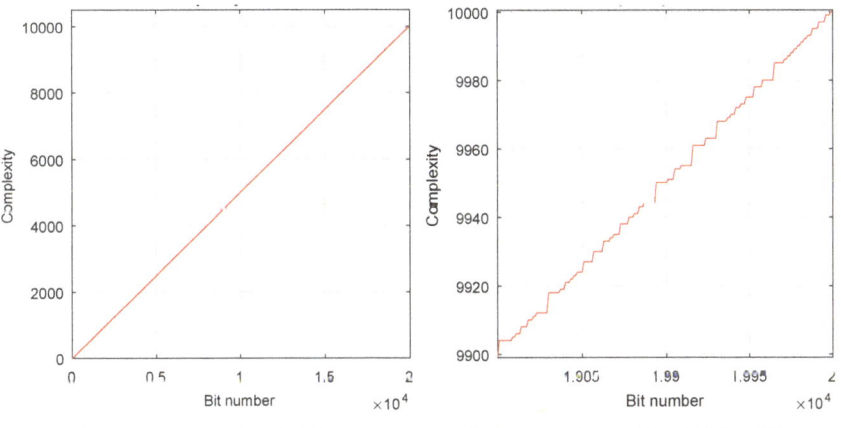

(a) Linear Complexity of $s(G)$.

(b) Zoom of the graphic of *LC* for $s(G)$.

Figure 4. Linear Complexity of $s(G)$ for the first 20,000 most significant bits.

3. Shannon Entropy and Min-Entropy

The entropy of a sequence is defined as a measure of the amount of information of a process measured in bits or as a measure of the uncertainty of a random variable. From these two possible interpretation, the quality of the output sequence or the input of a random number generator can be described, respectively.

Shannon's entropy is measured based on the average probability of all the values that the variable can take. A formal definition can be presented as follows,

Definition 2. *Let X be a random variable that takes on the values $x_1, x_2, \ldots, x_n$. Then the Shannon's entropy is defined as*

$$H(X) = -\sum_{i=1}^{n} Pr(x_i) \cdot \log_2(Pr(x_i)),$$

where $Pr(\cdot)$ represents probability.

If the process is a sequence of integers modulo m perfectly random, then its entropy is equal to n. As in the case at hand $m = 2^n$, the entropy of a random sequence must be close to $n = 8$ bit per octet.

The min-Entropy is only measured based on the probability of the more frequent occurrence value of the variable. It is recommended by the NIST SP 800 – 90B standard for True Random Number Generators (TRNG).

In order to determine if the proposed generator is considered perfect from these entropies values, according to Reference [40] for a sequence of 2^{20} octets, it must obtain a Shannon entropy value greater or equal than 7.976 bits per octet and a min-entropy greater or equal to 7.91 bits per octet. In this case the following values are obtained:

Shannon entropy (measured) = 7.9999 bits per octet.
Min-entropy (measured) = 7.9457 bits per octet,

then, it can be considered that this generator is correct using entropies. Note that the Shannon's entropy value of 7.9999 bits per octet fits close to the theoretical perfection of 8 bits per octet.

4. Lyapunov exponent

Lyapunov exponent measures the rate of divergence of nearby trajectories, which is a key component of chaotic dynamics. It is used as a quantitative measure for the sensitive dependence on initial conditions. It is desirable that two very close initial conditions (for instance, seeds or keys) provide very different trajectories (sequences). If Lyapunov exponent is greater than zero, the distance between two close initial conditions rapidly increases in the time, which means there exists an exponential divergence of the trajectories of a chaotic system. This value gives an idea of how different are the sequences generated by similar seeds, a very important feature to avoid attacks on the key of the generator. So, Lyapunov exponent is, in this case, a useful tool to evaluate the key space.

Next, a formal definition of Lyapunov exponent [41] is given.

Definition 3. *Consider d_0 the measure of the initial distance between two sequences and d_t the measure of the distance between the same sequences but after t iterations. We define Lyapunov exponent as:*

$$LE = \frac{1}{t} \ln\left(\left|\frac{d_t}{d_0}\right|\right).$$

If $LE = 0$, the sequences decrease their distance, tend to join and confused in one. The system converges and it is not at all random. If $LE > 0$, the distance increases, there is dependence sensitive to initial conditions, there is an exponential divergence of the orbit and randomness grows as higher is the value of LE.

Note that the Lyapunov exponent uses the natural logarithm of the Euclidean distance. Nevertheless, in information theory, other type of distances for measuring the distance between two sequences are used, for example Hamming distance, which indicates the number of bit positions in which both sequences differ.

If the Lyapunov exponent is modified simply by using the Hamming distance instead of the logarithm of the Euclidean distance, then it is called the Lyapunov Hamming exponent (LHE). If two numbers are identical, then its LHE value will be 0. Nevertheless, if all the bits of both numbers are different, then its LHE will be $LHE = \log_2 m = \log_2 2^n = n$, where n is the number of bits with which the numbers are encoded.

Obtaining the Lyapunov Hamming exponent for the chosen sequence is done by calculating the average of the LHE between every two consecutive numbers of the sequence. The best value will be $n/2$.

For this case, the best value is 4; we show the value obtained for our particular sequence analyzed:
Lyapunov Hamming exponent, ideal = 4.
Lyapunov Hamming exponent, real = 4.
Absolute deviation from the ideal = -1.0014×10^{-5}.
hence, the proposed generator passes perfectly this test.

5. Samples in increasing order

The samples of 8 bits are ordered by increasing value and are represented by a graph. They should give a continuous straight line (red), with an inclination of 45 degrees, which must cover the blue reference line.

This representation means that all the numbers are generated (if it is continuous) and that the density is uniform (if its inclination is 45 degrees). In Figure 5a, we observe that the samples are perfectly represented by a continuous straight line with the perfect inclination of 45 degrees.

From Figure 5b, the deviation between the increasing samples is analysed and the values $-1, 0$ or 1 are obtained.

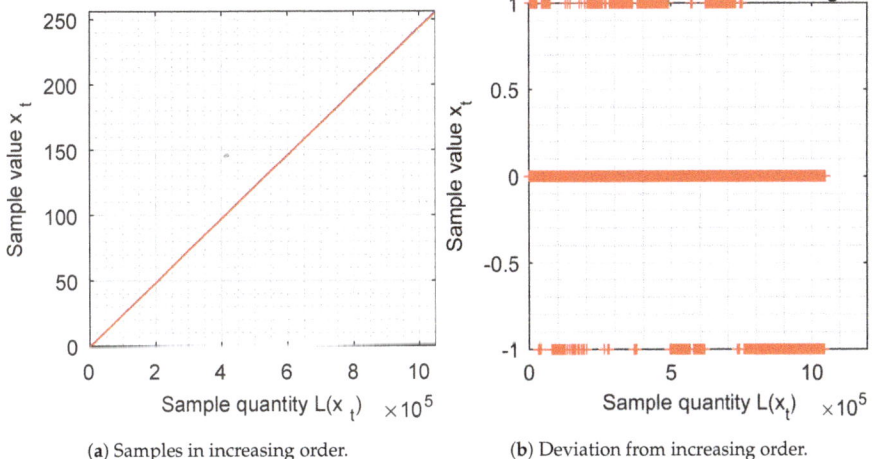

(a) Samples in increasing order. (b) Deviation from increasing order.

Figure 5. Samples ordered by increasing value.

6. Chaos game

Chaos game is a method that allows converting a one-dimensional sequence into a two dimensions sequence providing a very provocative visual representation, which reveals some of the statistical properties of the sequence under study. From this graphical technique is easy to look for, visually, patterns in the sequences generated by a random number generator. Furthermore, it allows us to find non-randomness within pseudo-random sequences.

Chaos game can be described mathematically by an Iterated Functions System (IFS) [10,42,43] and through which the transition to chaos associated with fractals can be studied. The result of chaos game is called attractor and not always is a fractal, it may be any compact set. If the output is a graph with fractals or patterns, then it means that the sequence cannot be considered random.

In Figure 6, it cannot be observe any pattern or fractal, it is a messy (or unordered) cloud of points, which does not provide any useful information for analysis, which implies good randomness.

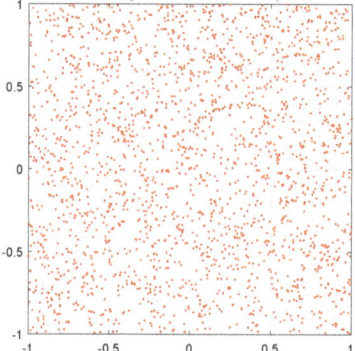

Figure 6. GSS generator Chaos game.

In order to better understand this graphical test, we present in Figure 7a,b two Chaos Game representations, which appeared in Reference [10], which are not cryptographically secure. Their graphics are fractal which indicates that the design depends on a pattern (denoting the lack of randomness) and it is also worth mentioning that this pattern could be used to obtain important information for cryptanalysis.

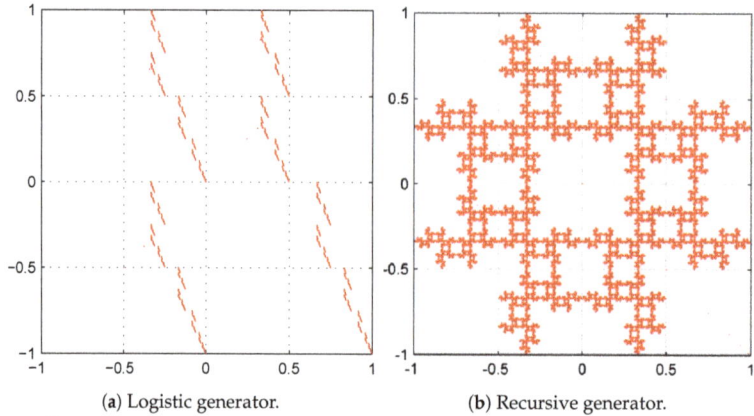

(**a**) Logistic generator. (**b**) Recursive generator.

Figure 7. Chaos game representations of imperfect generators. The observed patterns indicate a lack of randomness in the sequence.

7. Autocorrelation

The analysis of autocorrelation is a mathematical tool for finding repeating patterns analysing different sections of a message and compares them to find similarities. The autocorrelation function is defined as the crosscorrelation of the sequence with itself and allows measuring the linear relationship between random variables of processes separated a certain distance. It is very useful for finding periodic patterns within a signal.

Figure 8 represents the autocorrelation index of our GSS-sequence, for all samples available. It can be seen that the sequence has a very long period, larger than the size of the sequence analyzed since the repetition frequency is not reached in the graph.

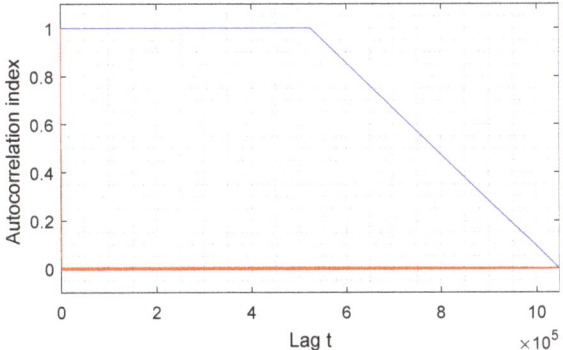

Figure 8. Autocorrelation function of a GSS-sequence.

The first autocorrelation coefficient is always equal to 1, while the other coefficients must have the smallest possible amplitude so that the sequence can be considered random before finding the period in which it begins to repeat itself. In the case at hand, values close to 0 are obtained, which means that the proposed sequence can be considered random for this study.

8. Fast Fourier Transform

The goal of Fast Fourier Transform test is the peak heights in the discrete Fast Fourier Transform. It consists of detecting repetitive patterns in the sequence analysed which would indicate a deviation from the assumption of randomness [8].

If the sequence is random, then all the maximum harmonics of Fast Fourier Transform have approximately the same horizontal level without an up or down trend.

Figure 9 shows that all amplitude values are included in the same range, which means that the test is passed.

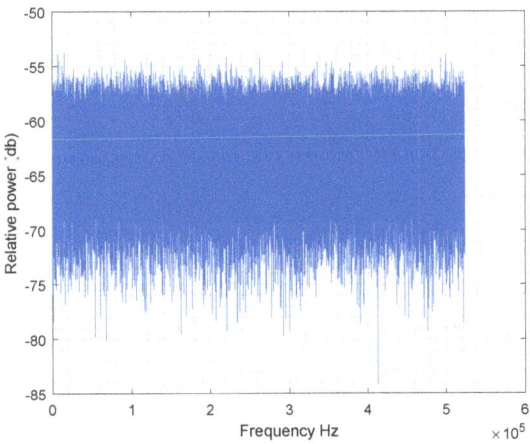

Figure 9. Fast Fourier Transform of s(G).

9. Distribution of identical samples

In this subsection, the distance of occurrence between samples of equal value is studied, because this measure is an important property of random sequences. The most probable distance between two identical samples of a perfect sequence is zero. If this distance increases, then the probability of coincidence between the two identical samples decreases following a Poisson distribution.

Figure 10 shows that the distribution of samples of the proposed sequence is close to the ideal.

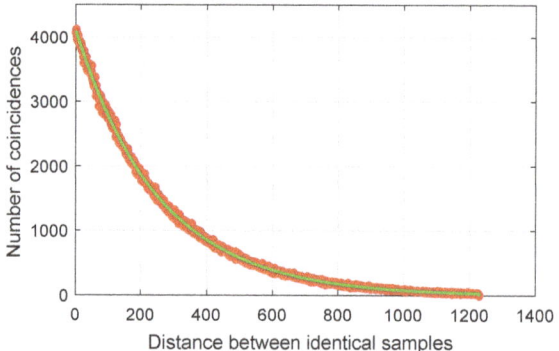

Figure 10. Distribution of samples with equal values a function of their distance: GSS-sequence (red) and a perfect random sequence (green).

10. Collisions of the sequence

Collisions are an intrinsic property of random sequences. If one has a sequence of integers module m, the amount of different integer numbers will be m. When a number appears repeated, we say that a collision has occurred. In Reference [44] an analysis of the collisions problem is presented based on the *birthday paradox* which states that in a group of k people chosen at random, at least a pair of them will have the same birthday with probability:

$$p_k = 1 - \left(1 - \frac{1}{m}\right)\left(1 - \frac{2}{m}\right)\cdots\left(1 - \frac{k-1}{m}\right), \qquad (3)$$

where m is the number of days of the year and k is the number of people in the living room.

This paradox can be applied to hash functions. One of the desirable properties of cryptographic hash functions is that it is computationally impossible for a collision to occur; that is, given two different inputs, hash function does not produce the same output.

Suppose that we have a hash function of n bits, so we have $m = 2^n$ output possible values. From this idea, it can be deduced the inequality:

$$k \geq \frac{1}{2} + \sqrt{2m \ln(2) + \frac{1}{4}},$$

which provides an estimated value of the quantity k of rolls of a random sequence that must be extracted to have a probability of a first collision greater than or equal to 0.5.

From Equation (3) it can be deduced the collision probability density distribution Dp_k as a function of k,

$$Dp_k = \frac{k-1}{m}\left(1 - \frac{1}{m}\right)\left(1 - \frac{2}{m}\right)\left(1 - \frac{3}{m}\right)\cdots\left(1 - \frac{k-2}{m}\right). \qquad (4)$$

In Figure 11 is represented the first collision probability density distribution function for a sequence of octets, that is, $n = 8$, $m = 256$ as a red line. It can be seen that the mode of the distribution is $k = 17 = 1 + \sqrt{m}$ and for a quantity of rolls $k = 4\sqrt{m} = 64$ the collision probability density is practically zero.

Any sequence with a perfect randomness must fit the first collision probability density distribution function corresponding to Equation (4).

The Figure 11 represents also a bar graph, with one bar for each value of k, of a GSS-sequence of 2^{20} octets. It can be seen the perfect fitting with the expected theoretical distribution.

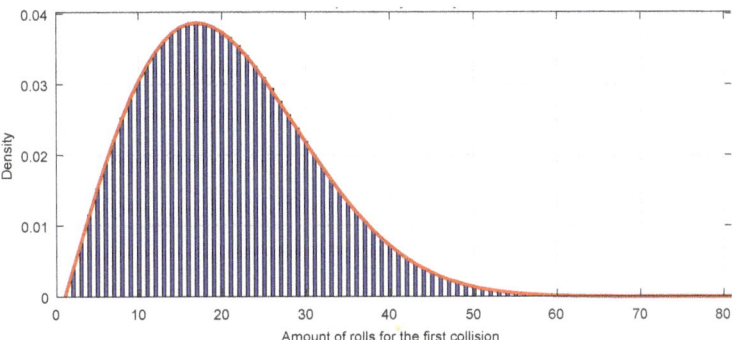

Figure 11. Distribution of the first collisions (blue bars) and collision probability density distribution function (red line).

As a curiosity, the first collision probability density distribution function coincides with a Weibull distribution function for the variable k, that is, the distribution which is most used to model data from reliability against catastrophes; in the present case, it models the amount of random number generation rolls needed for a first collision to appear, which is also a catastrophe for a hash function.

5.2. Diehard Battery of Tests

Diehard battery of tests [7] is a reliable standard and a powerful instrument for practical evaluation of the randomness of sequences of pseudo-random number generators. This tool is the first step in the evaluation process of cryptographic primitives. It cannot guarantee if your generator can be considered perfectly random, but if it does not pass the test suite, then it is not suitable for cryptographic applications.

Diehard battery consists of 15 different independent statistical tests, some of them repeated but with different parameters. The Diehard tests employ chi-squared goodness-to-fit technique to calculate a p-value, which should be uniform on $[0,1)$ if the input file contains truly independent random bits. It is considered that a bit stream really fails when it is gotten p-values of 0 or 1 to six or more places.

The GSS-sequences with characteristic polynomial of degree ≤ 27 have passed all tests in the Diehard battery. In Table 8 we show the results obtained with the Diehard battery from a $s(G)$ sequence with characteristic polynomial $p(x) = x^{27} + x^{23} + x^{22} + x^{17} + 1$.

Table 8. Diehard battery of tests results for a GSS sequence with characteristic polynomial of degree 27.

Test Name	p-Value	Result	Test Name	p-Value	Result
	0.854161			0.6612	
	0.128374			0.1300	
	0.350541			0.7321	
	0.843946			0.7540	
Birthday spacing	0.820384	Pass		0.7276	
	0.751627			0.0776	
	0.669644			0.2807	
	0.263248			0.2276	
	0.274206			0.5481	
Overlapping	0.973492	Pass		0.0144	
permutations	0.998474			0.7242	
	0.460374			0.7410	
Binary ranks	0.607801	Pass		0.6259	
	0.470376			0.5815	
	0.59389		OQSO	0.3380	Pass
	0.95088			0.8546	
	0.84285			0.5279	
	0.99576			0.3305	
	0.91144			0.1022	
	0.06885			0.3367	
	0.69611			0.8353	
	0.28168			0.6487	
	0.60022			0.5748	
Bit stream	0.93126	Pass		0.8688	
(Monkey tests)	0.77314			0.2946	
	0.91404			0.4309	
	0.81248			0.8943	
	0.60022			0.1388	

Table 8. Cont.

Test Name	p-Value	Result	Test Name	p-Value	Result
	0.84285			0.6424	
	0.94645			0.1627	
	0.96610			0.5008	
	0.83486			0.6695	
	0.52578			0.2392	
	0.99599			0.7181	
	0.9170			0.5722	
	0.9852			0.9521	
	0.6537			0.9762	
	0.3155			0.3309	
	0.2258			0.9433	
	0.9600			0.2852	
	0.6056			0.7472	
	0.9116			0.3780	
	0.7067			0.4109	
	0.8025			0.8180	
	0.9201			0.3395	
OPSO	0.9671	Pass		0.2346	
	0.2808		DNA	0.5149	Pass
	0.5257			0.9901	
	0.8779			0.0708	
	0.9751			0.0209	
	0.9980			0.9450	
	0.3569			0.9835	
	0.1756			0.2135	
	0.8006			0.0099	
	0.9974			0.9157	
	0.4474			0.0761	
	0.9458			0.9593	
Count-the-1's	0.923369	Pass		0.1119	
(stream of bytes)	0.375390			0.5837	
	0.069242		Parking lot	0.357527	Pass
	0.453489		Minimum distance	0.752286	Pass
	0.531694		3D Spheres	0.947691	Pass
	0.476337		Squeeze	0.990622	Pass
	0.115181		Overlapping sums	0.276467	Pass
	0.238283			0.276783	
	0.248038		Runs	0.893007	Pass
	0.170200			0.908305	
	0.595302			0.913183	

Table 8. Cont.

Test Name	p-Value	Result	Test Name	p-Value	Result
	0.167417		Craps	0.995956	Pass
	0.574701			105661	
Count-the-1's	0.384873	Pass			
(specific bytes)	0.944743				
	0.955924				
	0.210026				
	0.142320				
	0.717744				
	0.191102				
	0.728247				
	0.297792				
	0.971290				
	0.323464				
	0.408101				
	0.013264				
	0.859849				

5.3. FIPS Test 140-2. Security Requirements for Cryptographic Modules

FIPS (Federal Information Processing Standard) Publication 140-2, is a U.S. government computer security standard [6] used to approve cryptographic modules. The National Institute of Standards and Technology (NIST) issued the FIPS 140-2 publication series to coordinate the requirements and standards for cryptography modules that include both hardware and software components (last updated 2002).

In FIPS 140-2 there are 4 statistical random number generator tests—The Monobit Test, The Poker Test, The Runs Test and The Long Runs Test. The proposed GSS-sequences with characteristic polynomials of degree ≤ 27 pass all these tests. Below we detail the results:

1. LONG RUNS TEST(PRS): Passed. There are no runs of more than 25 equal bits.
2. MONOBIT TEST(PRS): Passed. The test is passed if $(9725 <$ number of ones $< 10275)$. Our result was: 9954.
3. X= POKER TEST(PRS): Passed. The test is passed if $2.16 < X < 46.17$;. Our result was: $X = 10.0736$.
4. RUNS TEST(PRS): Passed. The test is passed if the runs (for both the runs of zeros, red line, and the runs of ones, blue line) that occur (of lengths 1 through 6) are each within the corresponding interval specified in the Figure 12 by the green line.

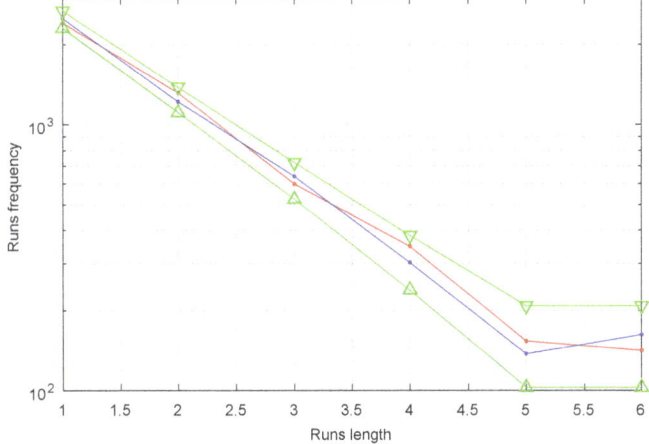

Figure 12. Run test for a GSS-sequence with characteristic polynomials of degree ≤ 27. Observe that the test is passed both for the runs of zeros (red line) and for the runs of ones (blue line) since they all fall within the corresponding range specified by the green line.

5.4. Lempel-Ziv Compression Test

The goal of this test is the number of cumulatively distinct patterns in the sequence. This test consists of determining how much is possible to compress the analysed sequence. If the sequence can be significantly compressed, it is considered to be non-random. The proposed GSS-sequences with characteristic polynomials of degree ≤ 27, pass this test with perfect results.

As can be seen throughout this section, the analyzed generator meets all the requirements needed to be used in the field of cryptography, according to points 1–4 mentioned in Section 1. Further work would be to study the resistance of this generator against the cryptographic attacks reported in the literature (Section 1, point 5).

6. Conclusions

In this article, we have found a relationship between two families of binary sequences belong to the class of decimation-based sequence generators, that is, the t-modified self-shrunken sequences can be generated from a generalized self-shrinking generator. We have analysed this relationship from two different points of view—one of them as binary sequences and other using the cyclotomic cosets. Furthermore, we have considered one of the most complete statistical test batteries for the study of randomness of sequences generated by the GSSG. In addition, we have reviewed some important graphical tests and basic and recent individual randomness tests found in the cryptographic literature. From the study of the last section, we can conclude that our random number generator (GSSG) produces good pseudo-random sequences since all the family of the sequences generated with characteristic polynomials of degree less than or equal to 27 pass satisfactorily the most important batteries of tests. The obtained results confirm the potential use of the generalized self-shrunken sequences for cryptographic purposes.

With regard to future work on this subject, the concatenation of GSS sequences from different primitive polynomials of different degrees could be analysed and studied, as well as the resistance of this generator against cryptographic attacks reported in the literature. Another important future work would be to do a comparative study of our generator with other well-known generators used in cryptographic applications nowadays.

Author Contributions: All the authors have equally contributed to the reported research in conceptualization, methodology, software and manuscript revision.

Funding: This research received no external funding.

Acknowledgments: This research has been partially supported by Ministerio de Economía, Industria y Competitividad (MINECO), Agencia Estatal de Investigación (AEI), and Fondo Europeo de Desarrollo Regional (FEDER, UE) under project COPCIS, reference TIN2017-84844-C2-1-R, and by Comunidad de Madrid (Spain) under project CYNAMON (P2018/TCS-4566), also co-funded by FSE and European Union FEDER funds. The first author was supported by CAPES (Brazil). The second author was partially supported by Spanish grant VIGROB-287 of the Universitat d'Alacant. We would like to thank Fausto Montoya for his help with the analysis of the sequences.

Conflicts of Interest: The authors declare no conflict of interest.

References

1. Bhowmick, A.; Sinha, N.; Arjunan, R.V.; Kishore, B. Permutation-substitution architecture based image encryption algorithm using middle square and RC4 PRNG. In Proceedings of the 2017 International Conference on Inventive Systems and Control (ICISC), Coimbatore, India, 19–20 January 2017; pp. 1–6.
2. Wortman, P.; Yan, W.; Chandy, J.; Tehranipoor, F. P2M-based security model: Security enhancement using combined puf and PRNG models for authenticating consumer electronic devices. *IET Comput. Digit. Tech.* **2018**, *12*, 289–296. [CrossRef]
3. Bikram, P.; Trivedi, G.; Jan, P.; Nemec, Z. Efficient PRNG design and implementation for various high throughput cryptographic and low power security applications. In Proceedings of the 2019 29th International Conference Radioelektronika (RADIOELEKTRONIKA), Pardubice, Czech Republic, 16–18 April 2019; pp. 1–6.
4. Moufek, H.; Guenda, K.; Gulliver, T.A. A new variant of the McEliece cryptosystem based on QC-LDPC and QC-MDPC codes. *IEEE Commun. Lett.* **2017**, *21*, 714–717. [CrossRef]
5. Gong, G.; Helleseth, T.; Kumar, P.V. Solomon W. Golomb–Mathematician, Engineer, and Pioneer. *IEEE Trans. Inf. Theory* **2018**, *64*, 2844–2857. [CrossRef]
6. FIPS PUB 140-2. Security Requirements for Cryptographic Modules. In *Federal Information Processing Standards Publication 140-2*; U.S. Department of Commerce, NIST, National Technical Information Service: Springfield, VA, USA, 2001.
7. Marsaglia, G. *The Marsaglia Random Number CDROM including the DIehard Battery of Tests of Randomness*; Florida State University: Tallahassee, FL, USA, 1995. Available online: http://www.stat.fsu.edu/pub/diehard (accessed on 3 November 2019).
8. National Institute of Standards and Technology. *A Statistical Test Suite for Random and Pseudorandom Number Generators for Cryptographic Applications*; NIST800-22, SP 800-22Rev 1a; 2010, U.S. Department of Commerce: Gaithersburg, MD, USA.
9. Orúe López, A.B. Contribución al Estudio del Criptoanálisis y Diseño de los Criptosistemas Caóticos. Ph.D. Thesis, Universidad Politécnica de Madrid, Escuela Técnica Superior de Ingenieros de Telecomunicación, Madrid, Spain, 2013.
10. Orúe, A.B.; Fúster-Sabater, A.; Fernández, V.; Montoya, F.; Hernández, L.; Martín, A. Herramientas gráficas de la criptografía caótica para el análisis de la calidad de secuencias pseudoaleatorias. In Proceedings of the Actas de la XIV Reunión Española sobre Criptología y Seguridad de la Información, RECSI XIV; Menorca, Illes Balears, Spain, 26–28 October 2016; pp. 180–185.
11. Hu, Y.; Xiao, G. Generalized self-shrinking generator. *IEEE Trans. Inf. Theory* **2004**, *50*, 714–719. [CrossRef]
12. Klein, A. Linear Feedback Shift Registers. In *Stream Ciphers*; Springer: London, UK, 2013; Chapter 2, pp. 1–13.
13. Coppersmith, D.; Krawczyk, H.; Mansour, Y. The shrinking generator. In Proceedings of the 13th Annual International Cryptology Conference on Advances in Cryptology (CRYPTO '93), Santa Barbara, CA, USA, 22–26 August 1993; Springer: Berlin/Heidelberg, Germany, 1994; pp. 22–39.
14. Meier, W.; Staffelbach, O. The self-shrinking generator. In *Advances in Cryptology, Proceedings of EUROCRYPT 1994*; Cachin, C., Camenisch, J., Eds.; Lecture Notes in Computer Science; Springer, Berlin/Heidelberg, Germany, 1984; Volume 950, pp. 205–214.
15. Dong, L.; Zeng, Y.; Hu, Y. F-gss: A novel fcsr-based keystream generator. In Proceedings of the First International Conference on Information Science and Engineering, Nanjing, China, 26–28 December 2009; pp. 1737–1740.
16. Kanso, A. Modified self-shrinking generator. *Comput. Electr. Eng.* **2010**, *36*, 993–1001. doi:10.1016/j.compeleceng.2010.02.004. [CrossRef]

17. Berzina, I.; Bets, R.; Buls, J.; Cers, E.; Kulesa, L. On a non-periodic shrinking generator. In Proceedings of the 2011 13th International Symposium on Symbolic and Numeric Algorithms for Scientific Computing, Timisoara, Romania, 26–29 September 2011; pp. 348–354.
18. Tasheva, A.T.; Tasheva, Z.N.; Milev, A.P. Generalization of the self-shrinking generator in the Galois Field GF(p^n). *Adv. Artif. Intell.* **2011**, *2011*, doi:10.1155/2011/464971. [CrossRef]
19. Tasheva, A.; Nakov, O.; Tasheva, Z. About balance property of the p-ary generalized self-shrinking generator sequence. In Proceedings of the 14th International Conference on Computer Systems and Technologies (CompSysTech '13), Ruse, Bulgaria, 28–29 June 2013; ACM: New York, NY, USA, 2013; pp. 299–306.
20. Erkek, E.; Tuncer, T. The implementation of asg and sg random number generators. In Proceedings of the 2013 International Conference on System Science and Engineering (ICSSE), Budapest, Hungary, 4–6 July 2013; pp. 363–367.
21. Boztas, S.; Alamer, A. Statistical dependencies in the self-shrinking generator. In Proceedings of the 2015 Seventh International Workshop on Signal Design and its Applications in Communications (IWSDA), Bengaluru, India, 14–18 September 2015; pp. 42–46.
22. Savova-Tasheva, Z.; Tasheva, A. Analysis of keystream produced by generalized shrinking multiplexing generator controlled by ternary m-sequence. In Proceedings of the 9th Balkan Conference on Informatics (BCI'19), Sofia, Bulgaria, 26–28 September 2019; ACM: New York, NY, USA, 2019; pp. 1–7.
23. Gergely, A.M.; Crainicu, B. A succinct survey on (pseudo-)random number generators from a cryptographic perspective. In Proceedings of the 2017 5th International Symposium on Digital Forensic and Security (ISDFS), Tirgu Mures, Romania, 26–28 April 2017; pp. 1–6.
24. Bikram, P.; Khobragade, A.; Sai, S.; Goswami, S.S.P.; Dutt, S.; Trivedi, G. Design and implementation of low-power high-throughput PRNGs for security applications. In Proceedings of the 2019 32nd International Conference on VLSI Design and 2019 18th International Conference on Embedded Systems, Delhi, NCR, India, 5–9 January 2019; pp. 535–536.
25. Prokofiev, A.O.; Chirkin, A.V.; Bukharov, V.A. Methodology for quality evaluation of PRNG, by investigating distribution in a multidimensional space. In Proceedings of the 2018 IEEE Conference of Russian Young Researchers in Electrical and Electronic Engineering (EIConRus), Moscow, Russia, 29 January–1 February 2018; pp. 355–357.
26. Dalai, D.K.; Maitra, S.; Pal, S.; Roy, D. Distinguisher and non-randomness of grain-v1 for 112, 114 and 116 initialisation rounds with multiple-bit difference in ivs. *IET Inf. Secur.* **2019**, *13*, 603–613. [CrossRef]
27. Avaroglu, E.; Çavdar, T. Quantum random number generators. In Proceedings of the 2018 International Conference on Artificial Intelligence and Data Processing (IDAP), Malatya, Turkey, 28–30 September 2018; pp. 1–4.
28. Zhu, S.; Ma, Y.; Li, X.; Yang, J.; Lin, J.; Jing, J. On the analysis and improvement of min-entropy estimation on time-varying data. *IEEE Trans. Inf. Forensics Secur.* **2019**, *1*, doi:10.1109/TIFS.2019.2947871. [CrossRef]
29. Liu, Y.; Tong, X. Hyperchaotic system-based pseudorandom number generator. *IET Inf. Secur.* **2016**, *10*, 433–441. [CrossRef]
30. Cardell, S.D.; Fúster-Sabater, A. Discrete linear models for the self-shrunken sequences. *Finite Fields Their Appl.* **2017**, *47*, 222–241. [CrossRef]
31. Cardell, S.D.; Fúster-Sabater, A. The t-Modified Self-shrinking Generator. In *Computational Science—ICCS 2018, Proceedings of the International Conference on Computational Science (ICCS 2018), Wuxi, China, 11–13 June 2018*; Shi, Y., Fu, H., Tian, Y., Krzhizhanovskaya, V.V., Lees, M.H., Dongarra, J., Sloot, P.M.A., Eds.; Lecture Notes in Computer Science; Springer: Cham, Switzerland, 2018; Volume 10860, pp. 653–663.
32. Golomb, S.W. *Shift Register-Sequences*; Aegean Park Press: Laguna Hill, CA, USA, 1982.
33. Fúster-Sabater, A. Linear Solutions for Irregularly Decimated Generators of Cryptographic Sequences. *Int. J. Nonlinear Sci. Numer. Simul.* **2014**, *15*, 377–385.[CrossRef]
34. Todorova, M.; Stoyanov, B.; Szczypiorski, K.; Kordov, K. SHAH: Hash Function based on Irregularly Decimated Chaotic Map. *Int. J. Electron. Telecommun.* **2018**, *64*, 457–465, doi:10.24425/123546. [CrossRef]
35. Lidl, R.; Niederreiter, H. *Introduction to Finite Fields and Their Applications*; Cambridge University Press: New York, NY, USA, 1986.
36. Alvarez, G.; Montoya, F.; Romera, M.; Pastor, G. Cryptanalyzing an improved security modulated chaotic encryption scheme using ciphertext absolute value. *Chaos Soliton. Fract.* **2005**, *23*, 1749–1756. [CrossRef]
37. Paar, C.; Pelzl, J. *Understanding Cryptography*; Springer: Berlin/Heidelberg, Germany, 2010.

38. Massey, J.L. Shift-register synthesis and BCH decoding. *IEEE Trans. Inf. Theor.* **1969**, *15*, 122–127. doi:10.1109/TIT.1969.1054260. [CrossRef]
39. Rueppel, R.A. Linear Complexity and Random Sequences. In *Advances in Cryptology—EUROCRYPT 1985*; Pichler, F., Ed.; Lecture Notes in Computer Science; Springer: Berlin/Heidelberg, Germany, 1986; Volume 219, pp. 167–188.
40. Killmann, W.; Schindler, W. *AIS 20/AIS 31, A Proposal for: Functionality Classes for Random Number Generators*; Bundesamt für Sicherheit in der Informationstechnik (BSI): Frankfurt am Main, Germany, 2011.
41. Romera, M. *Técnica de los Sistemas Dinámicos Discretos*; 27 CSIC, Madrid II-C; Textos Universitarios: Madrid, Spain, 1997.
42. Peitgen, H.O.; Jurgens, H.; Saupe, D. *Chaos and Fractals: New Frontiers of Science*; Springer: New York, NY, USA, 2004.
43. Barnsley, M. *Fractals Everywhere*, 2nd ed.; Dover Publications, Inc.: Mineola, NY, USA, 2012.
44. Cormen, T.H.; Leiserson, C.E.; Rivest, R.L.; Stein, C. *Introduction to Algorithms*, 2nd ed.; The MIT Press: Cambridge, MA, USA, 2001.

© 2019 by the authors. Licensee MDPI, Basel, Switzerland. This article is an open access article distributed under the terms and conditions of the Creative Commons Attribution (CC BY) license (http://creativecommons.org/licenses/by/4.0/).

Article

The Symmetric Key Equation for Reed–Solomon Codes and a New Perspective on the Berlekamp–Massey Algorithm

Maria Bras-Amorós [1,*] and Michael E. O'Sullivan [2]

1. Departament d'Enginyeria Informàtica i Matemàtiques, Universitat Rovira i Virgili, Av. Països Catalans 26, 43007 Tarragona, Catalonia
2. Department of Mathematics and Statistics, San Diego State University, 5500 Campanile Drive, San Diego, CA 92182-7720, USA; mosullivan@sdsu.edu
* Correspondence: maria.bras@urv.cat

Received: 19 September 2019; Accepted: 25 October 2019; Published: 2 November 2019

Abstract: This paper presents a new way to view the key equation for decoding Reed–Solomon codes that unites the two algorithms used in solving it—the Berlekamp–Massey algorithm and the Euclidean algorithm. A new key equation for Reed–Solomon codes is derived for simultaneous errors and erasures decoding using the symmetry between polynomials and their reciprocals as well as the symmetries between dual and primal codes. The new key equation is simpler since it involves only degree bounds rather than modular computations. We show how to solve it using the Euclidean algorithm. We then show that by reorganizing the Euclidean algorithm applied to the new key equation we obtain the Berlekamp–Massey algorithm.

Keywords: Reed–Solomon codes; key equation; Berlekamp–Massey algorithm; Sugiyama et al. algorithm; euclidean algorithm

1. Introduction

Reed–Solomon codes are the basis of many applications such as secret sharing [1], distributed storage [2,3], private information retrieval [4] and the analysis of cryptographic hardness [5]. The most used tool for decoding Reed–Solomon codes is the key equation by Berlekamp [6] and the milestone algorithms that solve it are the Berlekamp–Massey algorithm [7] and the Sugiyama et al. adaptation of the Euclidean algorithm [8]. Their connections are analyzed in [9–12]. This paper is meant to bring a new unified presentation of the key equation, the Sugiyama-Euclidean algorithm and the Berlekamp–Massey algorithm for correcting errors and erasures for Reed–Solomon codes.

Section 2 presents a revisited key equation for both erasures and errors using the symmetry between polynomials and their reciprocals as well as the symmetries between dual and primal codes. In the new key equation, as opposed to the classical equation, there is no need to reference computations modulo a power of the indeterminate, and the correction polynomials reveal error locations rather than their inverses. Section 3 gives a way to solve the new key equation using the Euclidean algorithm. We show how the Berlekamp–Massey algorithm can be obtained by reorganizing the Euclidean algorithm. Hence, the whole paper is, in fact, a simple presentation of the Berlekamp–Massey algorithm as a restructured Euclidean algorithm.

2. Symmetric Key Equation

2.1. Reed–Solomon Codes

Suppose that $\mathbb{F}$ is a finite field of q elements and suppose that α is a primitive element of $\mathbb{F}$. Let $n = q - 1$. Each vector $u = (u_0, \ldots, u_{n-1}) \in \mathbb{F}^n$ is identified with the polynomial $u(x) = u_0 + u_1 x + \cdots + u_{n-1} x^{n-1}$. The evaluation of $u(x)$ at a is then denoted $u(a)$. The cyclic code $C^*(k)$ of length n generated by the polynomial $(x - \alpha)(x - \alpha^2) \cdots (x - \alpha^{n-k})$ is classically referred to as a (primal) Reed–Solomon code. Its dimension is k. On the other hand, the cyclic code $C(k)$ of lenth n generated by the polynomial $(x - \alpha^{n-(k+1)})(x - \alpha^{n-(k+2)}) \cdots (x - \alpha)(x - 1)$ is referred to as a dual Reed–Solomon code. Its dimension is k as well. The minimum distance of both codes is $d = n - k + 1$. The codes are related by the equality $C(k)^\perp = C^*(n - k)$.

The vector space $\mathbb{F}^n$ is naturally bijected to itself through a map $c \mapsto c^*$ taking $C(k)$ to $C^*(k)$. For a vector $c = (c_0, c_1, \ldots, c_{n-1}) \in \mathbb{F}^n$ the vector c^* is defined componentwise as $c^* = (c_0, \alpha^{-1} c_1, \alpha^{-2} c_2, \ldots, \alpha c_{n-1})$. Symmetrically, if $c^* = (c_0^*, c_1^*, \ldots, c_{n-1}^*)$, then $c = (c_0^*, \alpha c_1^*, \alpha^2 c_2^*, \ldots, \alpha^{n-1} c_{n-1}^*)$. In particular, $c(\alpha^i) = c_0^* + \alpha c_1^* \alpha^i + \alpha^2 c_2^* \alpha^{2i} + \cdots + \alpha^{n-1} c_{n-1}^* \alpha^{(n-1)i} = c^*(\alpha^{i+1})$.

Due to this bijective map, algorithms for correcting errors and erasures for primal Reed–Solomon code are also applicable for dual Reed–Solomon codes and vice versa. Indeed, if the codeword $c \in C(k)$ at minimum distance of a received vector u differs from u by a vector of errors e, then the codeword $c^* \in C^*(k)$ at minimum distance of a received vector u^* differs from u^* by a vector of errors e^*.

2.2. Decoding for Errors and Erasures

Suppose that a noisy channel adds t errors and erases s other components of a transmitted codeword $c \in C(k)$ with $2t + s < d$. Let u be the received word after replacing the erased positions by 0 and let $e = u - c$. The *erasure locator polynomial* is defined as $\Lambda_r = \prod_{i : c_i \text{ was erased}} (x - \alpha^i)$ while the *error locator polynomial* is defined as $\Lambda_e = \prod_{i : e_i \neq 0, c_i \text{ not erased}} (x - \alpha^i)$. The product $\Lambda_r \Lambda_e$ is called Λ. We remark that while Λ_r is known driectly from the received word, the Λ_e is not *a priori* known. The *error evaluator polynomial* is defined as $\Omega = \sum_{i : e_i \neq 0 \text{ or } c_i \text{ erased}} e_i \prod_{j : e_j \neq 0 \text{ or } c_j \text{ erased and } j \neq i} (x - \alpha^i) = \sum_{i=0}^{n-1} e_i \frac{\Lambda}{x - \alpha^i}$ The error positions can be identified by $\Lambda_e(\alpha^i) = 0$ while the error values can be derived, as well as the erased values, from the analogue of the Forney formula [13]

$$e_i = \frac{\Omega(\alpha^i)}{\Lambda'(\alpha^i)}.$$

Notice that in the traditional setting, the roots of the locator polynomial are not related to the error positions but to their inverses. Hence, in the new setting we take the reciprocals of the polynomials of the traditional setting thus establishing a symmetry between the different versions. Also, the classical Forney formula involves the evaluator polynomial and the derivative of the locator polynomial evaluated at the inverses of the error positions, while with the new settings we use directly the error positions.

Finally, the polynomial $S = e(\alpha^{n-1}) + e(\alpha^{n-2})x + \cdots + e(\alpha)x^{n-2} + e(1)x^{n-1}$ is called the *syndrome polynomial* of e.

Lemma 1. $\Omega(x^n - 1) = \Lambda S$.

Proof. We can compute directly,

$$\begin{aligned}
\Omega(x^n - 1) &= (x^n - 1) \sum_{i=0}^{n-1} e_i \frac{\Lambda}{x - \alpha^i} \\
&= \Lambda \sum_{i=0}^{n-1} e_i \frac{x^n - 1}{x - \alpha^i} \\
&= \Lambda \sum_{i=0}^{n-1} e_i \sum_{j=0}^{n-1} x^{n-1-j} (\alpha^i)^j \\
&= \Lambda \sum_{j=0}^{n-1} x^{n-1-j} \sum_{i=0}^{n-1} e_i (\alpha^j)^i \\
&= \Lambda \sum_{j=0}^{n-1} x^{n-1-j} e(\alpha^j) \\
&= \Lambda S
\end{aligned}$$

□

The general term of S is $e(\alpha^{n-1-i})x^i$, but we only know from a received word the values $e(1) = u(1), \ldots, e(\alpha^{n-k-1}) = u(\alpha^{n-k-1})$. For this reason we use the *truncated syndrome polynomial* defined as $\tilde{S} = e(\alpha^{n-k-1})x^k + e(\alpha^{n-k-2})x^{k+1} + \cdots + e(1)x^{n-1}$. The degree of the polynomial $\Omega(x^n - 1) - \Lambda\tilde{S} = \Lambda(S - \tilde{S})$ is at most $t + s + k - 1 < \frac{d-s}{2} + s + n - d = n - \frac{d-s}{2}$. One consequence of this bound is that the reciprocal polynomials $\Omega^* = x^{t+s-1}\Omega(1/x)$, $\Lambda^* = x^{t+s}\Lambda(1/x)$ and the polynomial $\tilde{S}^* = x^{n-1}\tilde{S}(1/x)$ satisfy the well known Berlekamp key equation $\Lambda^*\tilde{S}^* \equiv \Omega^* \mod x^{n-s-k}$. Theorem 1 uses the bound on the degree of $\Omega(x^n - 1) - \Lambda\tilde{S}$ to derive a symmetric key equation for dual Reed–Solomon codes. To prove it, we first need the next two lemmas.

Lemma 2. *Suppose that f is a polynomial of $\mathbb{F}[x]$ with $\deg(f) < n$. Suppose that for a given $\alpha \in \mathbb{F}^*$ the polynomial $f(x)\frac{x^n - 1}{x - \alpha}$ has no term of degree $n - 1$. Then α is a root of f.*

Proof. The Euclidean division of f by $x - \alpha$ gives a polynomial $g \in \mathbb{F}[x]$ of degree smaller than $n - 1$ that satisfies $f(x) = f(\alpha) + g(x)(x - \alpha)$. Then $f(x)\frac{x^n - 1}{x - \alpha} = f(\alpha)\frac{x^n - 1}{x - \alpha} + g(x)(x^n - 1)$. On one hand, the product $g(x)(x^n - 1)$ has no term of degree $n - 1$. On the other hand, the coefficient of $f(\alpha)\frac{x^n - 1}{x - \alpha}$ of degree $n - 1$ is exactly $f(\alpha)$. Hence, if $f(x)\frac{x^n - 1}{x - \alpha}$ has no term of degree $n - 1$, then necessarily $f(\alpha) = 0$. □

Lemma 3. *Suppose that f is a polynomial of $\mathbb{F}[x]$ with $\deg(f) \leq n - s - t$ such that the terms of degree $n - t, \ldots, n - 1$ of $f\Lambda_r S$ are all zero. Then Λ_e is a divisor of f.*

Proof. Suppose that the terms of degree $n-t,\ldots,n-1$ of $f\Lambda_r S$ are all zero. Suppose c_j was not erased and $e_j \neq 0$. Consider $g(x) = \Lambda_e/(x-\alpha_j)$. We have $\deg(g) = t-1$ and consequently the term of degree $n-1$ of $fg\Lambda_r S$ is 0. Then,

$$\begin{aligned}
fg\Lambda_r S &= f(x)g(x)\Lambda_r(x)\frac{\Omega(x)(x^n-1)}{\Lambda(x)} \\
&= \sum_{k:e_k \neq 0} e_k f(x)g(x)\Lambda_r(x)\frac{x^n-1}{x-\alpha_k} \\
&= e_j f(x)g(x)\Lambda_r(x)\frac{x^n-1}{x-\alpha_j} \\
&\quad + \sum_{\substack{k:e_k \neq 0, \\ c_k \text{ not erased} \\ k \neq j}} e_k f(x)\frac{g(x)}{x-\alpha_k}\Lambda_r(x)(x^n-1) \\
&\quad + \sum_{k:c_k \text{ erased}} e_k f(x)g(x)\frac{\Lambda_r(x)}{x-\alpha_k}(x^n-1).
\end{aligned}$$

Because of the restriction on the degree of f, none of the last two summands has term of degree $n-1$. Since the term of degree $n-1$ of $fg\Lambda_r S$ is 0, so is the term of degree $n-1$ of $f(x)g(x)\Lambda_r(x)\frac{x^n-1}{x-\alpha_j}$. By Lemma 2, $x-\alpha_j$ must be a divisor of f. Since j was chosen arbitrarily such that $e_j \neq 0$ and c_j was not erased, we conclude that Λ_e must divide f. □

Theorem 1 (Symmetric key equation). *Suppose that a number s of erasures occurred together with a number of at most $\lfloor \frac{d-s-1}{2} \rfloor$ errors. Then the polynomials Λ_e and Ω are uniquely determined by the conditions*

1. f is monic
2. f, φ are coprime
3. $\deg(f) \leq \frac{d-s}{2}$
4. $\deg(f\Lambda_r \tilde{S} - \varphi(x^n-1)) < n - \frac{d-s}{2}$

Proof. It is easy to see that Λ_e and Ω satisfy conditions 1, 2, 3. It follows from the previous lemmas that Λ_e and Ω satisfy condition 4. Conversely, suppose that f, φ satisfy the conditions 3 and 4. We will prove that the terms of degrees $n-t,\ldots,n-1$ of $f\Lambda_r S$ are all zero. Then, by Lemma 3, and because $\deg(f) \leq \frac{d-s}{2} \leq n - \frac{d+s}{2} = n-s-\frac{d-s}{2} < n-s-t$, it can be deduced that Λ_e is a divisor of f. Indeed, write

$$f\Lambda_r S = (f\Lambda_r \tilde{S} - \varphi(x^n-1)) + f\Lambda_r(S-\tilde{S}) + \varphi(x^n-1).$$

By consition 4, the degree of the first term in this sum is less than $n - \frac{d-s}{2} < n-t$. By condition 3, $\deg(f\Lambda_r(S-\tilde{S})) \leq \frac{d-s}{2} + s + k - 1 = n - \frac{d-s}{2} < n-t$. By condition 4, $\deg(\varphi) + n \leq \deg(f) + s + n - 1$. Consequently $\deg(\varphi) < \deg(f) + s$ and by condition 3, $\deg(\varphi) < \frac{d-s}{2} + s = \frac{d+s}{2} \leq n - \frac{d-s}{2} < n-t$. So, the terms of degrees $n-t,\ldots,n-1$ of $\varphi(x^n-1)$ are all zero. Suppose now that there exists $g \in \mathbb{F}[x]$ such that $f = g\Lambda_e$. Then

$$\begin{aligned}
f\Lambda_r\tilde{S} - \varphi(x^n-1) &= f\Lambda_r(\tilde{S}-S) + f\Lambda_r S - \varphi(x^n-1) \\
&= f\Lambda_r(\tilde{S}-S) + g\Lambda S - \varphi(x^n-1) \\
&= f\Lambda_r(\tilde{S}-S) + g\Omega(x^n-1) - \varphi(x^n-1) \\
&= f\Lambda_r(\tilde{S}-S) + (g\Omega - \varphi)(x^n-1).
\end{aligned}$$

By condition 4, $\deg(f\Lambda_r\tilde{S} - \varphi(x^n-1)) < n - \frac{d-s}{2}$ and as just seen, $\deg(f\Lambda_r(\tilde{S}-S)) < n-t$. Consequently, $\varphi = g\Omega$. Now condition 1 and condition 2 imply $g = 1$ and so $\varphi = \Omega$ and $f = \Lambda_e$. □

3. Solving the Symmetric Key Equation

We first approach the case in which only erasures occurred. In this case $\Lambda = \Lambda_r$, $\Lambda_e = 1$, and Ω can be directly derived from the key equation of Theorem 1. Indeed, the polynomial Ω is exactly the sum of those monomials of $\Lambda_r \bar{S}$ of degree at least $n - \frac{d-s}{2}$, divided by the monomial $x^{n - \frac{d-s}{2}}$.

Suppose now the case in which errors and erasures occured simultaneously. The extended Euclidean algorithm applied to the quotient polynomial $\Lambda_r \bar{S}$ and the divisor polynomial $-(x^n - 1)$ gives $\gcd(\Lambda_r \bar{S}, x^n - 1)$ and two polynomials $\lambda(x)$ and $\eta(x)$ satisfying that $\lambda \Lambda_r \bar{S} - \eta(x^n - 1) = \gcd(\Lambda_r \bar{S}, x^n - 1)$. A new remainder r_i and two polynomials $\lambda_i(x)$ and $\eta_i(x)$ such that $\lambda_i \Lambda_r \bar{S} - \eta_i(x^n - 1) = r_i$ are computed at each intermediate step of the Euclidean algorithm, in a way such that the degree of r_i decreases at each step. Truncating at a proper point the Euclidean algorithm we can obtain two polynomials λ_i and η_i satisfying that the degree of $\lambda_i \Lambda_r \bar{S} - \eta_i(x^n - 1)$ is smaller than $n - \frac{d-s}{2}$. The next algorithm is a truncated version of the Euclidean algorithm. It satisfies that, for all $i \geq 0$, $\deg(r_i) \leq \deg(r_{i-1})$ and $\deg(f_i) \geq \deg(f_{i-1})$.

Algorithm 1: Euclidean Algorithm

Initialize:
$r_{-2} = \Lambda_r \bar{S}, \quad f_{-2} = 1, \quad \varphi_{-2} = 0,$
$r_{-1} = -(x^n - 1), \quad f_{-1} = 0, \quad \varphi_{-1} = 1,$

while $\deg(r_i) \geq n - \frac{d-s}{2}$:

$q_i = \text{Quotient}(r_{i-2}, r_{i-1})$
$r_i = \text{Remainder}(r_{i-2}, r_{i-1})$
$f_i = f_{i-2} - q_i f_{i-1}$
$\varphi_i = \varphi_{i-2} - q_i \varphi_{i-1}$

end while
Return $f_i / \text{LC}(f_i), \varphi_i / \text{LC}(f_i)$

or, equivalently, in matrix form,

Initialize:
$$\begin{pmatrix} r_{-1} & f_{-1} & \varphi_{-1} \\ r_{-2} & f_{-2} & \varphi_{-2} \end{pmatrix} = \begin{pmatrix} -(x^n - 1) & 0 & 1 \\ \Lambda_r \bar{S} & 1 & 0 \end{pmatrix}$$

while $\deg(r_i) \geq n - \frac{d-s}{2}$:

$q_i = \text{Quotient}(r_{i-2}, r_{i-1})$

$$\begin{pmatrix} r_i & f_i & \varphi_i \\ r_{i-1} & f_{i-1} & \varphi_{i-1} \end{pmatrix} = \begin{pmatrix} -q_i & 1 \\ 1 & 0 \end{pmatrix} \begin{pmatrix} r_{i-1} & f_{i-1} & \varphi_{i-1} \\ r_{i-2} & f_{i-2} & \varphi_{i-2} \end{pmatrix}$$

end while
Return $f_i / \text{LC}(f_i), \varphi_i / \text{LC}(f_i)$

For every integer i larger than or equal to -1 consider the matrix $\begin{pmatrix} \mathring{R}_i & \mathring{F}_i & \mathring{\Phi}_i \\ \mathring{R}_i & \mathring{F}_i & \mathring{\Phi}_i \end{pmatrix} =$
$\begin{pmatrix} 1/\text{LC}(r_i) & 0 \\ 0 & -\text{LC}(r_i) \end{pmatrix} \begin{pmatrix} r_i & f_i & \varphi_i \\ r_{i-1} & f_{i-1} & \varphi_{i-1} \end{pmatrix}$ It is easy to check that the polynomial $\mathring{R}_i$ is monic. In the algorithm one can replace the update step by the next multiplication.

$$\begin{pmatrix} \mathring{R}_i & \mathring{F}_i & \mathring{\Phi}_i \\ \mathring{R}_i & \mathring{F}_i & \mathring{\Phi}_i \end{pmatrix} = \begin{pmatrix} \frac{1}{\text{LC}(\mathring{R}_{i-1} - Q_i \mathring{R}_{i-1})} & 0 \\ 0 & -\text{LC}(\mathring{R}_{i-1} - Q_i \mathring{R}_{i-1}) \end{pmatrix} \begin{pmatrix} -Q_i & 1 \\ 1 & 0 \end{pmatrix} \begin{pmatrix} \mathring{R}_{i-1} & \mathring{F}_{i-1} & \mathring{\Phi}_{i-1} \\ \mathring{R}_{i-1} & \mathring{F}_{i-1} & \mathring{\Phi}_{i-1} \end{pmatrix},$$

where the polynomial Q_i is the quotient of the division of $\mathring{R}_{i-1}$ by $\mathring{R}_{i-1}$. Furthermore, if $Q_i = Q_i^{(0)} + Q_i^{(1)} x + \cdots + Q_i^{(l_i)} x^{l_i}$, then

$$\begin{pmatrix} -Q_i & 1 \\ 1 & 0 \end{pmatrix} = \begin{pmatrix} 1 & -Q_i^{(0)} \\ 0 & 1 \end{pmatrix} \begin{pmatrix} 1 & -Q_i^{(1)}x \\ 0 & 1 \end{pmatrix} \cdots \begin{pmatrix} 1 & -Q_i^{(l)}x^l \\ 0 & 1 \end{pmatrix} \begin{pmatrix} 0 & 1 \\ 1 & 0 \end{pmatrix}$$ and the update step becomes

$$\begin{pmatrix} \mathring{R}_i & \mathring{F}_i & \mathring{\Phi}_i \\ \mathring{R}_i & \mathring{F}_i & \mathring{\Phi}_i \end{pmatrix} = \begin{pmatrix} \frac{1}{LC(\mathring{R}_{i-1}-Q_i\mathring{R}_{i-1})} & 0 \\ 0 & -LC(\mathring{R}_{i-1}-Q_i\mathring{R}_{i-1}) \end{pmatrix} \begin{pmatrix} 1 & -Q_i^{(0)} \\ 0 & 1 \end{pmatrix} \begin{pmatrix} 1 & -Q_i^{(1)}x \\ 0 & 1 \end{pmatrix} \cdots$$

$$\cdots \begin{pmatrix} 1 & -Q_i^{(l)}x^l \\ 0 & 1 \end{pmatrix} \begin{pmatrix} 0 & 1 \\ 1 & 0 \end{pmatrix} \begin{pmatrix} \mathring{R}_{i-1} & \mathring{F}_{i-1} & \mathring{\Phi}_{i-1} \\ \mathring{R}_{i-1} & \mathring{F}_{i-1} & \mathring{\Phi}_{i-1} \end{pmatrix},$$

One can see that $LC(\mathring{R}_{i-1} - Q_i\mathring{R}_{i-1})$ and the $Q_i^{(j)}$'s are the leading coefficients of the left-most, top-most polynomials in the previous product of all the previous matrices. This follows from the fact that $\mathring{R}_i$ is monic. Define μ as the (changing) leading coefficients of the left-most, top-most element in the product of all the previous matrices. It follows that

$$\begin{pmatrix} \mathring{R}_i & \mathring{F}_i & \mathring{\Phi}_i \\ \mathring{R}_i & \mathring{F}_i & \mathring{\Phi}_i \end{pmatrix} = \begin{pmatrix} \frac{1}{\mu} & 0 \\ 0 & -\mu \end{pmatrix} \begin{pmatrix} 1 & -\mu \\ 0 & 1 \end{pmatrix} \begin{pmatrix} 1 & -\mu x \\ 0 & 1 \end{pmatrix} \cdots \begin{pmatrix} 1 & -\mu x^{l_i} \\ 0 & 1 \end{pmatrix} \begin{pmatrix} 0 & 1 \\ 1 & 0 \end{pmatrix} \begin{pmatrix} \mathring{R}_{i-1} & \mathring{F}_{i-1} & \mathring{\Phi}_{i-1} \\ \mathring{R}_{i-1} & \mathring{F}_{i-1} & \mathring{\Phi}_{i-1} \end{pmatrix}$$

$$= \begin{pmatrix} \frac{1}{\mu} & 0 \\ 0 & -\mu \end{pmatrix} \begin{pmatrix} 1 & -\mu \\ 0 & 1 \end{pmatrix} \begin{pmatrix} 1 & -\mu x \\ 0 & 1 \end{pmatrix} \cdots \begin{pmatrix} 1 & -\mu x^{l_i} \\ 0 & 1 \end{pmatrix} \begin{pmatrix} 0 & 1 \\ 1 & 0 \end{pmatrix}$$
$$\begin{pmatrix} \frac{1}{\mu} & 0 \\ 0 & -\mu \end{pmatrix} \begin{pmatrix} 1 & -\mu \\ 0 & 1 \end{pmatrix} \begin{pmatrix} 1 & -\mu x \\ 0 & 1 \end{pmatrix} \cdots \begin{pmatrix} 1 & -\mu x^{l_{i-1}} \\ 0 & 1 \end{pmatrix} \begin{pmatrix} 0 & 1 \\ 1 & 0 \end{pmatrix} \begin{pmatrix} \mathring{R}_{i-2} & \mathring{F}_{i-2} & \mathring{\Phi}_{i-2} \\ \mathring{R}_{i-2} & \mathring{F}_{i-2} & \mathring{\Phi}_{i-2} \end{pmatrix}$$

$$= \begin{pmatrix} \frac{1}{\mu} & 0 \\ 0 & -\mu \end{pmatrix} \begin{pmatrix} 1 & -\mu \\ 0 & 1 \end{pmatrix} \begin{pmatrix} 1 & -\mu x \\ 0 & 1 \end{pmatrix} \cdots \begin{pmatrix} 1 & -\mu x^{l_i} \\ 0 & 1 \end{pmatrix} \begin{pmatrix} 0 & -\mu \\ 1/\mu & 0 \end{pmatrix}$$
$$\begin{pmatrix} 1 & -\mu \\ 0 & 1 \end{pmatrix} \begin{pmatrix} 1 & -\mu x \\ 0 & 1 \end{pmatrix} \cdots \begin{pmatrix} 1 & -\mu x^{l_{i-1}} \\ 0 & 1 \end{pmatrix} \begin{pmatrix} 0 & 1 \\ 1 & 0 \end{pmatrix} \begin{pmatrix} \mathring{R}_{i-2} & \mathring{F}_{i-2} & \mathring{\Phi}_{i-2} \\ \mathring{R}_{i-2} & \mathring{F}_{i-2} & \mathring{\Phi}_{i-2} \end{pmatrix}$$

$$= \begin{pmatrix} \frac{1}{\mu} & 0 \\ 0 & -\mu \end{pmatrix} \begin{pmatrix} 1 & -\mu \\ 0 & 1 \end{pmatrix} \begin{pmatrix} 1 & -\mu x \\ 0 & 1 \end{pmatrix} \cdots \begin{pmatrix} 1 & -\mu x^{l_i} \\ 0 & 1 \end{pmatrix} \begin{pmatrix} 0 & -\mu \\ 1/\mu & 0 \end{pmatrix}$$
$$\begin{pmatrix} 1 & -\mu \\ 0 & 1 \end{pmatrix} \begin{pmatrix} 1 & -\mu x \\ 0 & 1 \end{pmatrix} \cdots \begin{pmatrix} 1 & -\mu x^{l_{i-1}} \\ 0 & 1 \end{pmatrix} \begin{pmatrix} 0 & -\mu \\ 1/\mu & 0 \end{pmatrix}$$
$$\vdots$$
$$\begin{pmatrix} 1 & -\mu \\ 0 & 1 \end{pmatrix} \begin{pmatrix} 1 & -\mu x \\ 0 & 1 \end{pmatrix} \cdots \begin{pmatrix} 1 & -\mu x^{l_0} \\ 0 & 1 \end{pmatrix} \begin{pmatrix} 0 & 1 \\ 1 & 0 \end{pmatrix} \begin{pmatrix} \mathring{R}_{-1} & \mathring{F}_{-1} & \mathring{\Phi}_{-1} \\ \mathring{R}_{-1} & \mathring{F}_{-1} & \mathring{\Phi}_{-1} \end{pmatrix},$$

Let us label the matrices in the previous product:

$$\begin{pmatrix} \frac{1}{\mu} & 0 \\ 0 & -\mu \end{pmatrix} \overbrace{\begin{pmatrix} 1 & -\mu \\ 0 & 1 \end{pmatrix}}^{M_m} \overbrace{\begin{pmatrix} 1 & -\mu x \\ 0 & 1 \end{pmatrix}}^{M_{m-1}} \cdots \begin{pmatrix} 1 & -\mu x^{l_i-1} \\ 0 & 1 \end{pmatrix} \begin{pmatrix} 1 & -\mu x^{l_i} \\ 0 & 1 \end{pmatrix} \begin{pmatrix} 0 & -\mu \\ 1/\mu & 0 \end{pmatrix}$$
$$\vdots$$
$$\begin{pmatrix} 1 & -\mu \\ 0 & 1 \end{pmatrix} \begin{pmatrix} 1 & -\mu x \\ 0 & 1 \end{pmatrix} \cdots \overbrace{\begin{pmatrix} 1 & -\mu x^{l_1-1} \\ 0 & 1 \end{pmatrix}}^{M_{l_0+3}} \overbrace{\begin{pmatrix} 1 & -\mu x^{l_1} \\ 0 & 1 \end{pmatrix}}^{M_{l_0+2}} \overbrace{\begin{pmatrix} 0 & -\mu \\ 1/\mu & 0 \end{pmatrix}}^{M_{l_0+1}}$$
$$\overbrace{\begin{pmatrix} 1 & -\mu \\ 0 & 1 \end{pmatrix}}^{M_{l_0}} \overbrace{\begin{pmatrix} 1 & -\mu x \\ 0 & 1 \end{pmatrix}}^{M_{l_0-1}} \cdots \overbrace{\begin{pmatrix} 1 & -\mu x^{l_0-1} \\ 0 & 1 \end{pmatrix}}^{M_1} \overbrace{\begin{pmatrix} 1 & -\mu x^{l_0} \\ 0 & 1 \end{pmatrix}}^{M_0} \begin{pmatrix} 0 & 1 \\ 1 & 0 \end{pmatrix} \cdot \begin{pmatrix} \mathring{R}_{-1} & \mathring{F}_{-1} & \mathring{\Phi}_{-1} \\ \mathring{R}_{-1} & \mathring{F}_{-1} & \mathring{\Phi}_{-1} \end{pmatrix}$$

Now, we define

$$\begin{pmatrix} R_{-1} & F_{-1} & \Phi_{-1} \\ \tilde{R}_{-1} & \tilde{F}_{-1} & \tilde{\Phi}_{-1} \end{pmatrix} = \begin{pmatrix} 0 & 1 \\ 1 & 0 \end{pmatrix} \begin{pmatrix} \mathring{R}_{-1} & \mathring{F}_{-1} & \mathring{\Phi}_{-1} \\ \mathring{R}_{-1} & \mathring{F}_{-1} & \mathring{\Phi}_{-1} \end{pmatrix} = \begin{pmatrix} \Lambda_r \bar{S} & 1 & 0 \\ x^n - 1 & 0 & -1 \end{pmatrix}$$

$$\begin{pmatrix} R_i & F_i & \Phi_i \\ \tilde{R}_i & \tilde{F}_i & \tilde{\Phi}_i \end{pmatrix} = M_i \cdot M_{i-1} \cdots M_0 \cdot \begin{pmatrix} R_{-1} & F_{-1} & \Phi_{-1} \\ \tilde{R}_{-1} & \tilde{F}_{-1} & \tilde{\Phi}_{-1} \end{pmatrix}$$

Lets us see now that, for all $i \leq m$, the polynomials $\tilde{R}_i$ and F_i are monic. Indeed, $\tilde{R}_{-1} = x^n - 1$ is monic, and it follows by induction and by the definition of the matrices M_i, that $\tilde{R}_i$ is monic for all i. Now, all the matrices M_i have determinant equal to 1. This implies that $R_i \tilde{F}_i - F_i \tilde{R}_i$ is constant for all i and it equals $-(x^n - 1)$. In particular, since $LC(R_i \tilde{F}_i - F_i \tilde{R}_i) = -LC(F_i)LC(\tilde{R}_i) = -LC(F_i)$, we deduce that for every i, the polynomial F_i is monic.

Algorithm 2 computes the matrices $\begin{pmatrix} R_i & F_i & \Phi_i \\ \tilde{R}_i & \tilde{F}_i & \tilde{\Phi}_i \end{pmatrix}$ until $\deg(R_i) < n - \frac{d-s}{2}$.

Algorithm 2: Single Coefficient Euclidean Algorithm.

Initialize:

$$\begin{pmatrix} R_{-1} & F_{-1} & \Phi_{-1} \\ \tilde{R}_{-1} & \tilde{F}_{-1} & \tilde{\Phi}_{-1} \end{pmatrix} = \begin{pmatrix} \Lambda_r \bar{S} & 1 & 0 \\ x^n - 1 & 0 & -1 \end{pmatrix}$$

while $\deg(R_i) \geq n - \frac{d-s}{2}$:

$\mu = \mathbf{LC}(R_i)$
$p = \mathbf{deg}(R_i) - \mathbf{deg}(\tilde{R}_i)$

if $p \geq 0$ **then**

$$\begin{pmatrix} R_{i+1} & F_{i+1} & \Phi_{i+1} \\ \tilde{R}_{i+1} & \tilde{F}_{i+1} & \tilde{\Phi}_{i+1} \end{pmatrix} = \begin{pmatrix} 1 & -\mu x^p \\ 0 & 1 \end{pmatrix} \begin{pmatrix} R_i & F_i & \Phi_i \\ \tilde{R}_i & \tilde{F}_i & \tilde{\Phi}_i \end{pmatrix}$$

else

$$\begin{pmatrix} R_{i+1} & F_{i+1} & \Phi_{i+1} \\ \tilde{R}_{i+1} & \tilde{F}_{i+1} & \tilde{\Phi}_{i+1} \end{pmatrix} = \begin{pmatrix} 0 & -\mu \\ 1/\mu & 0 \end{pmatrix} \begin{pmatrix} R_i & F_i & \Phi_i \\ \tilde{R}_i & \tilde{F}_i & \tilde{\Phi}_i \end{pmatrix}$$

end if

end while
Return F_i, Φ_i

Due to the fact that the polynomials $\tilde{R}_i$ are monic, after each step with a negative value of p the new updated value p coincides with the previous one but with opposite sign and so happens for μ. Taking this into account we join each step with a negative value of p with the next step. We obtain

$$\begin{pmatrix} R_{i+1} & F_{i+1} & \Phi_{i+1} \\ \tilde{R}_{i+1} & \tilde{F}_{i+1} & \tilde{\Phi}_{i+1} \end{pmatrix} = \begin{pmatrix} 1 & \mu x^{-p} \\ 0 & 1 \end{pmatrix} \begin{pmatrix} 0 & -\mu \\ 1/\mu & 0 \end{pmatrix} \begin{pmatrix} R_i & F_i & \Phi_i \\ \tilde{R}_i & \tilde{F}_i & \tilde{\Phi}_i \end{pmatrix}$$

This adjustment keeps F_i, Φ_i unaltered. It can be stated as follows

At this point we observe that we only need to keep the polynomials R_i (and $\tilde{R}_i$) because we need their leading coefficients (the μ_i's). The next lemma proves that these leading coefficients can be obtained independently of the polynomials R_i. This allows the computation of the polynomials F_i, Φ_i iteratively while dispensing with the polynomials R_i.

Lemma 4. $LC(R_i) = LC(F_i \Lambda_r \bar{S})$

Proof. The result is obvious for $i = -1$. Since we joined two steps, before Algorithm 3, the degree of the remainder $R_i = F_i \Lambda_r \tilde{S} - \Phi_i(x^n - 1) = F_i \Lambda_r \tilde{S} - x^n \Phi_i + \Phi_i$ is at most $n - 1$ for every $i \geq 1$. Consequently all terms of $x^n \Phi_i$ cancel with terms of $F_i \Lambda_r \tilde{S}$ and R_i must have leading term equal to either a term of Φ_i or a term of $F_i \Lambda_r \tilde{S}$ or a sum of a term of Φ_i and a term of $F_i \Lambda_r \tilde{S}$.

On the other hand, the algorithm computes $\mathrm{LC}(R_i)$ only while $\deg(R_i) \geq n - \frac{d-s}{2}$. In particular, $2\deg(R_i) = 2n - d + s \geq n + s$. Leu us show that in this case the degree of the leading term of R_i is strictly larger than the degree of Φ_i. Indeed, since all the matrices M_i in the algorithm have determinant equal to 1, this implies that $\deg(\Phi_i) = \deg(\Lambda_r \tilde{S}) - \deg(\tilde{R}_i) \leq n + s - \deg(\tilde{R}_i) < 2\deg(R_i) - \deg(R_i) = \deg(R_i)$. □

Algorithm 3: Refactored Single Coefficient Euclidean Algorithm

Initialize:

$$\begin{pmatrix} R_{-1} & F_{-1} & \Phi_{-1} \\ \tilde{R}_{-1} & \tilde{F}_{-1} & \tilde{\Phi}_{-1} \end{pmatrix} = \begin{pmatrix} \Lambda_r \tilde{S} & 1 & 0 \\ x^n - 1 & 0 & -1 \end{pmatrix}$$

while $\deg(R_i) \geq n - \frac{d-s}{2}$:

$\mu = \mathrm{LC}(R_i)$
$p = \deg(R_i) - \deg(\tilde{R}_i)$
if $p \geq 0$ or $\mu = 0$ **then**

$$\begin{pmatrix} R_{i+1} & F_{i+1} & \Phi_{i+1} \\ \tilde{R}_{i+1} & \tilde{F}_{i+1} & \tilde{\Phi}_{i+1} \end{pmatrix} = \begin{pmatrix} 1 & -\mu x^p \\ 0 & 1 \end{pmatrix} \begin{pmatrix} R_i & F_i & \Phi_i \\ \tilde{R}_i & \tilde{F}_i & \tilde{\Phi}_i \end{pmatrix}$$

else

$$\begin{pmatrix} R_{i+1} & F_{i+1} & \Phi_{i+1} \\ \tilde{R}_{i+1} & \tilde{F}_{i+1} & \tilde{\Phi}_{i+1} \end{pmatrix} = \begin{pmatrix} x^{-p} & -\mu \\ 1/\mu & 0 \end{pmatrix} \begin{pmatrix} R_i & F_i & \Phi_i \\ \tilde{R}_i & \tilde{F}_i & \tilde{\Phi}_i \end{pmatrix}$$

end if

end while
Return F_i, Φ_i

We transform now Algorithm 3 in a way such that isntead of keeping the remainders we keep their degrees. For this we use the values $d_i, \tilde{d}_i$ satisfying, at each step, that $d_i \geq \deg(R_i), \tilde{d}_i = \deg(\tilde{R}_i)$.

Algorithm 4 is exactly the Berlekamp–Massey algorithm applied to the recurrence $\sum_{j=0}^{t} \Lambda_j e(\alpha^{i+j-1}) = 0$ for all $i > 0$. This linear recurrence is a consequence of the equality $\frac{S}{x^n - 1} = \frac{1}{x}\left(e(1) + \frac{e(\alpha)}{x} + \frac{e(\alpha^2)}{x^2} + \cdots\right)$ and the fact that $\Lambda \frac{S}{x^n - 1}$ is a polynomial and, hence, its terms of negative order in its expression as a Laurent series in $1/x$ are all zero.

Algorithm 4: Berlekamp-Massey Algorithm

Initialize:

$d_{-1} = s + \deg(\tilde{S})$
$\tilde{d}_{-1} = n$
$\begin{pmatrix} F_{-1} & \Phi_{-1} \\ \tilde{F}_{-1} & \tilde{\Phi}_{-1} \end{pmatrix} = \begin{pmatrix} 1 & 0 \\ 0 & -1 \end{pmatrix}$

while $d_i \geq n - \frac{d-s}{2}$:

$\mu = \mathbf{Coefficient}(F_i \Lambda_r \tilde{S}, d_i)$
$p = d_i - \tilde{d}_i$

if $p \geq 0$ or $\mu = 0$ then

$\begin{pmatrix} F_{i+1} & \Phi_{i+1} \\ \tilde{F}_{i+1} & \tilde{\Phi}_{i+1} \end{pmatrix} = \begin{pmatrix} 1 & -\mu x^p \\ 0 & 1 \end{pmatrix} \begin{pmatrix} F_i & \Phi_i \\ \tilde{F}_i & \tilde{\Phi}_i \end{pmatrix}$
$d_{i+1} = d_i - 1$
$\tilde{d}_{i+1} = \tilde{d}_i$

else

$\begin{pmatrix} F_{i+1} & \Phi_{i+1} \\ \tilde{F}_{i+1} & \tilde{\Phi}_{i+1} \end{pmatrix} = \begin{pmatrix} x^{-p} & -\mu \\ 1/\mu & 0 \end{pmatrix} \begin{pmatrix} F_i & \Phi_i \\ \tilde{F}_i & \tilde{\Phi}_i \end{pmatrix}$
$d_{i+1} = \tilde{d}_i - 1$
$\tilde{d}_{i+1} = d_i$

end if

end while

Return F_i, Φ_i

4. Conclusions

By working with error/erasure locator polynomials whose roots correspond to the error positions rather than to their inverses and with an evaluator polynomial that gives the error values when we evaluate it at the error positions instead of evaluating it at the inverses of the error positions we get to a symmetric key equation for Reed–Solomon codes. We showed that the symmetric key equation can be solved by an adapted Euclidean algorithm whose steps can be refined leading naturally to the Berlekamp–Massey algorithm.

Author Contributions: The authors contributed equally to the theoretical framing and algorithms and the corresponding author took principle responsibility for writing the article.

Funding: The first author was partly supported by the Catalan Government under grant 2017 SGR 00705, by the Spanish Ministry of Economy and Competitivity under grant TIN2016-80250-R, and by Universitat Rovira i Virgili under grant OPEN2019.

Conflicts of Interest: The authors declare no conflict of interest.

References

1. McEliece, R.J.; Sarwate, D.V. On sharing secrets and Reed-Solomon codes. *Commun. ACM* **1981**, *24*, 583–584. [CrossRef]
2. Dimakis, A.G.; Ramchandran, K.; Wu, Y.; Suh, C. A Survey on Network Codes for Distributed Storage. *Proc. IEEE* **2011**, *99*, 476–489. [CrossRef]

3. Tamo, I.; Ye, M.; Barg, A. The repair problem for Reed-Solomon codes: Optimal repair of single and multiple erasures with almost optimal node size. *IEEE Trans. Inf. Theory* **2019**, *65*, 2673–2695. [CrossRef]
4. Tajeddine, R.; Gnilke, O.W.; Karpuk, D.; Freij-Hollanti, R.; Hollanti, C. Private information retrieval from coded storage systems with colluding, Byzantine, and unresponsive servers. *IEEE Trans. Inf. Theory* **2019**, *65*, 3898–3906. [CrossRef]
5. Kiayias, A.; Yung, M. Cryptographic hardness based on the decoding of Reed-Solomon codes. *IEEE Trans. Inf. Theory* **2008**, *54*, 2752–2769. [CrossRef]
6. Berlekamp, E.R. *Algebraic Coding Theory*; McGraw-Hill Book Co.: New York, NY, USA, 1968; pp. xiv+466.
7. Massey, J.L. Shift-register synthesis and BCH decoding. *IEEE Trans. Inf. Theory* **1969**, *15*, 122–127. [CrossRef]
8. Sugiyama, Y.; Kasahara, M.; Hirasawa, S.; Namekawa, T. A method for solving key equation for decoding Goppa codes. *Inf. Control* **1975**, *27*, 87–99. [CrossRef]
9. Dornstetter, J.L. On the equivalence between Berlekamp's and Euclid's algorithms. *IEEE Trans. Inf. Theory* **1987**, *33*, 428–431. [CrossRef]
10. Heydtmann, A.E.; Jensen, J.M. On the equivalence of the Berlekamp-Massey and the Euclidean algorithms for decoding. *IEEE Trans. Inf. Theory* **2000**, *46*, 2614–2624.
11. Mateer, T.D. On the equivalence of the Berlekamp-Massey and the Euclidean algorithms for algebraic decoding. In Proceedings of the 12th Canadian Workshop on Information Theory (CWIT), Kelowna, BC, Canada, 17–20 May 2011; pp. 139–142.
12. Ilani, I. Berlekamp–Massey Algorithm: Euclid in Disguise. In Proceedings of the 2018 IEEE International Conference on the Science of Electrical Engineering in Israel (ICSEE), Eilat, Israel, 12–14 December 2018; pp. 1–5. [CrossRef]
13. Forney, G.D., Jr. On decoding BCH codes. *IEEE Trans. Inf. Theory* **1965**, *11*, 549–557. [CrossRef]

© 2019 by the authors. Licensee MDPI, Basel, Switzerland. This article is an open access article distributed under the terms and conditions of the Creative Commons Attribution (CC BY) license (http://creativecommons.org/licenses/by/4.0/).

Article
The Root Extraction Problem for Generic Braids

María Cumplido [1,2,*], Juan González-Meneses [3,*] and Marithania Silvero [4,*]

1. Institut de Mathématiques de Bourgogne, UMR 5584, CNRS, Univ. Bourgogne Franche-Comté, 21000 Dijon, France
2. Department of Mathematics, Heriot-Watt University, Edinburgh, Scotland EH14 4AS, UK
3. Departamento de Álgebra, Universidad de Sevilla, 41012 Sevilla, Spain
4. Departamento de Ciencias Integradas, Universidad de Huelva, 21007 Huelva, Spain
* Correspondence: M.Cumplido@hw.ac.uk (M.C.); meneses@us.es (J.G.-M.); marithania@us.es (M.S.)

Received: 24 September 2019; Accepted: 17 October 2019; Published: 23 October 2019

Abstract: We show that, generically, finding the k-th root of a braid is very fast. More precisely, we provide an algorithm which, given a braid x on n strands and canonical length l, and an integer $k > 1$, computes a k-th root of x, if it exists, or guarantees that such a root does not exist. The generic-case complexity of this algorithm is $O(l(l+n)n^3 \log n)$. The non-generic cases are treated using a previously known algorithm by Sang-Jin Lee. This algorithm uses the fact that the ultra summit set of a braid is, generically, very small and symmetric (through conjugation by the Garside element Δ), consisting of either a single orbit conjugated to itself by Δ or two orbits conjugated to each other by Δ.

Keywords: braid groups; algorithms in groups; group-based cryptography

1. Introduction

Group theory is 'the language of symmetry', as it is beautifully explained by Marcus du Sautoy in his book *Symmetry*. In this paper we will deal with a fascinating family of groups discovered by Emil Artin: Braid groups.

There are several computational problems in braid groups that have been proposed for their potential applications in cryptography [1]. Initially, the conjugacy problem in the braid group $\mathbb{B}_n$ was proposed as a non-commutative alternative to the discrete logarithm problem [2,3]. Later, some other problems were proposed, including the k-th root extraction problem: given $x \in \mathbb{B}_n$ and an integer $k > 1$, find $a \in \mathbb{B}_n$ such that $a^k = x$.

The interest of braid groups for cryptography has decreased considerably, mainly due to the appearance of algorithms which solve the conjugacy problem extremely fast in the generic case [4–6]. The main problem with the proposed cryptographic protocols turns out to be the key generation. Public and secret keys are chosen 'at random', and this implies that the protocols are insecure against algorithms which have a fast generic-case complexity.

While the future of braid-cryptography depends on finding a good key-generation procedure, there are some other problems in braid groups whose generic-case complexity is still to be studied. This is the case of the k-th root (extraction) problem.

A priori, the study of the generic case for the k-th root problem could be though to be nonsense as, generically, the k-th root of a braid x does not exist. But we should think of the braid x as the k-th power of a generic braid: in protocols based on this problem, a secret braid a is chosen at random, and the braid $x = a^k$ is made public. Hence we are dealing with braids for which a k-th root is known to exist. In any case, the algorithm in this paper not only shows that root extraction in braid groups is generically very fast, but can also be used by those mathematicians needing a simple algorithm for finding a k-th root of a braid (or proving that it does not exist), which works in most cases.

There are already known algorithms to solve the k-th root problem in braid groups and, more generally, in Garside groups [7,8]. But these algorithms can be simplified a lot in the generic case, as we will show in this paper.

The plan of this paper is as follows. In Section 2 we provide the necessary tools to describe the situation and attack the problem. Then in Section 3, we prove the theoretical results needed for our proposed algorithm, which is given in Section 4, together with the study of its generic-case complexity.

This generic-case complexity turns out to be quadratic on the canonical length l of the braid, if the number n of strands is fixed. More precisely, the generic-case complexity is $O(l(l+n)n^3 \log n)$ (Theorem 6).

2. Preliminaries

2.1. Garside Structure of $\mathbb{B}_n$

A group G is said to be a *Garside group* [9] if it admits a submonoid $\mathcal{P}$ (whose elements are called *positive*) such that $\mathcal{P} \cap \mathcal{P}^{-1} = \{1\}$, and a special element $\Delta \in \mathcal{P}$, called *Garside element*, satisfying the following properties:

- The partial order $\preccurlyeq$ in G defined by $a \preccurlyeq b$ if $a^{-1}b \in \mathcal{P}$ is a lattice order. If $a \preccurlyeq b$ we say that a is a *prefix* of b. The lattice structure implies that for all $a, b \in G$ there exists a unique meet $a \wedge b$ and a unique join $a \vee b$ with respect to $\preccurlyeq$. Notice that this partial order is invariant under left-multiplication.
- The set of simple elements $\mathcal{S} := \{s \in G \mid 1 \preccurlyeq s \preccurlyeq \Delta\}$ is finite and generates G.
- Conjugation by Δ preserves $\mathcal{P}$, that is, $\Delta^{-1}\mathcal{P}\Delta = \mathcal{P}$.
- $\mathcal{P}$ is atomic: the atoms are the indivisible elements of $\mathcal{P}$ (elements $a \in \mathcal{P}$ for which there is no decomposition $a = bc$ with non-trivial elements $b, c \in \mathcal{P}$). Then, for every $x \in \mathcal{P}$ there is an upper bound on the number of atoms in a decomposition of the form $x = a_1 a_2 \cdots a_n$, where each a_i is an atom.

One of the main examples of Garside groups is the braid group on n strands, denoted by $\mathbb{B}_n$. This group has a standard presentation due to Artin [10]:

$$\mathbb{B}_n = \left\langle \sigma_1, \sigma_2, ..., \sigma_{n-1} \;\middle|\; \begin{array}{ll} \sigma_i \sigma_j \sigma_i = \sigma_j \sigma_i \sigma_j & \text{if } |i-j| = 1 \\ \sigma_i \sigma_j = \sigma_j \sigma_i & \text{if } |i-j| > 1 \end{array} \right\rangle.$$

Attending to the above presentation, a braid is said to be positive if it can be written as a product of positive powers of the generators $\{\sigma_i\}_{i=1}^n$. The set of positive braids forms the monoid $\mathcal{P}$ corresponding to the classical Garside structure of $\mathbb{B}_n$. We will denote this monoid by $\mathbb{B}_n^+$.

The usual Garside element in $\mathbb{B}_n^+$, which we denote Δ_n, is defined recursively setting $\Delta_2 = \sigma_1$ and

$$\Delta_n = \Delta_{n-1} \sigma_{n-1} \sigma_{n-2} \cdots \sigma_1,$$

for all $n > 2$. We will often write Δ and omit the subindex n when there is no ambiguity.

Consider now the inner automorphism $\tau : \mathbb{B}_n \to \mathbb{B}_n$ determined by Δ. That is, $\tau(x) = \Delta^{-1} x \Delta$. One can easily show from the presentation of $\mathbb{B}_n$ that $\tau(\sigma_i) = \sigma_{n-i}$ for $1 \leq i \leq n-1$. Hence τ has order 2 and Δ^2 is central. In fact, the center of $\mathbb{B}_n$ is cyclic, generated by Δ^2 [11].

The set $\mathcal{S}$ of simple elements and the automorphism τ will be very important in the sequel.

2.2. Normal Forms, Cyclings and Decyclings

It is well-known that Garside groups have solvable word problem, as one can compute a normal form for each element.

Let us first define the *right complement* of a simple element $s \in \mathcal{S}$ as $\partial(s) = s^{-1}\Delta$. That is, $\partial(s)$ is the only element $t \in \mathcal{P}$ such that $st = \Delta$. Let us see that $\partial(s) = t$ is also a simple element. Recall that the simple elements are the positive prefixes of Δ. Since τ preserves $\mathcal{P}$ (by definition of Garside group), we have that $\tau(s)$ is positive. Now

$$st\tau(s) = \Delta\tau(s) = s\Delta,$$

hence $t\tau(s) = \Delta$, which implies that t is a positive prefix of Δ, that is, $t \in \mathcal{S}$. It follows that we have a map $\partial: \mathcal{S} \to \mathcal{S}$. Notice that, by definition, $\partial^2 \equiv \tau$.

Given two simple elements $s, t \in \mathcal{S}$, we say that the decomposition st is *left weighted* if s is the biggest possible simple element (with respect to $\preccurlyeq$) in any decomposition of the element st as a product of two simple elements. This condition can be restated as $\partial(s) \wedge t = 1$, i.e., $\partial(s)$ and t have no non-trivial prefixes in common.

Definition 1 ([12,13]). *The left normal form of an element $x \in \mathbb{B}_n$ is the unique decomposition $x = \Delta^p x_1 \cdots x_l$ so that $p \in \mathbb{Z}, l \geq 0, x_i \in \mathcal{S} \setminus \{1, \Delta\}$ for $i = 1, \ldots, l$, and $x_i x_{i+1}$ is a left weighted decomposition, for $i = 1, \ldots, l-1$.*

Given such a decomposition, we define the *infimum*, *supremum* and *canonical length* of x as $\inf(x) = p$, $\sup(x) = p + l$ and $\ell(x) = l$, respectively. Equivalently, the infimum and supremum of x can be defined as the maximum and minimum integers p and s so that $\Delta^p \preccurlyeq x \preccurlyeq \Delta^s$ (see [12]).

It is important to notice that conjugation by Δ preserves the Garside structure of $\mathbb{B}_n$. Hence, if the left normal form of a braid x is $\Delta^p x_1 \cdots x_l$, then the left normal form of $\tau(x)$ is $\Delta^p \tau(x_1) \cdots \tau(x_l)$. We will make use of this property later.

Garside groups also have solvable conjugacy problem. One of the main tools to solve problems related to conjugacy in braid groups are the summit sets, which are subsets of the conjugacy class of a braid. Throughout this article we are going to use two of them: the *super summit set* [12] and the *ultra summit set* [4]. Let us first introduce some concepts:

Definition 2. *Let $x = \Delta^p x_1 \cdots x_l$ be in left normal form, with $l > 0$. Notice that we can write:*

$$x = \tau^{-p}(x_1)\Delta^p x_2 \cdots x_l.$$

We define the initial factor of x as $\iota(x) = \tau^{-p}(x_1)$, and the final factor of x as $\varphi(x) = x_l$. We can then write:

$$x = \iota(x)\Delta^p x_2 \cdots x_l \quad \text{and} \quad x = \Delta^p x_1 \cdots x_{l-1} \varphi(x).$$

If $l = 0$, we set $\iota(x) = 1$ and $\varphi(x) = \Delta$.

Notice that, as τ^2 is the identity, we actually have either $\iota(x) = x_1$ if p is even, or $\iota(x) = \tau(x_1)$ if p is odd. This happens in braid groups, but not in other Garside groups in which the order of τ is bigger.

Definition 3 ([12]). *Let $x = \Delta^p x_1 \cdots x_l$ be in left normal form, with $l > 0$. The cycling and decycling of x are the conjugates of x defined, respectively, as*

$$\mathbf{c}(x) = \Delta^p x_2 \cdots x_l \iota(x) \quad \text{and} \quad \mathbf{d}(x) = \varphi(x)\Delta^p x_1 \cdots x_{l-1}.$$

Thus $\mathbf{c}(x)$ is the conjugate of x by $\iota(x)$, and that $\mathbf{d}(x)$ is the conjugate of x by $\varphi(x)^{-1}$.

Cyclings and decyclings were defined in [12] in order to try to simplify the braid x by conjugations. Usually, if $l \geq 2$, the decomposition $\Delta^p x_2 \cdots x_l \iota(x)$ is **not** the left normal form of $\mathbf{c}(x)$. So $\mathbf{c}(x)$ could a priori have a shorter normal form (with less factors). A similar situation happens for $\mathbf{d}(x)$.

If $\Delta^p x_2 \cdots x_l \iota(x)$ is actually the left normal form of $\mathbf{c}(x)$ (when $l \geq 2$), we say that the braid x is *rigid*. This happens if and only if $x_l \iota(x)$ (that is, $\varphi(x) \iota(x)$) is a left weighted decomposition. We can extend this definition to every case, when $l \geq 0$:

Definition 4. *We say that $x \in \mathbb{B}_n$ is rigid if $\varphi(x) \iota(x)$ is a left weighted decomposition.*

If x is rigid, neither cycling nor decycling can simplify its normal form $x = \Delta^p x_1 \cdots x_l$. Actually, the normal forms of the iterated cyclings of x are, if p is even:

$$\mathbf{c}(x) = \Delta^p x_2 \cdots x_l x_1, \qquad \mathbf{c}^2(x) = \Delta^p x_3 \cdots x_l x_1 x_2, \qquad \ldots$$

so $\mathbf{c}^l(x) = x$ in this case. In the case when p is odd we have:

$$\mathbf{c}(x) = \Delta^p x_2 \cdots x_l \tau(x_1), \qquad \mathbf{c}^2(x) = \Delta^p x_3 \cdots x_l \tau(x_1) \tau(x_2), \qquad \ldots$$

so $\mathbf{c}^{2l}(x) = x$ in this case.

In the same way, if x is rigid we have, for p even:

$$\mathbf{d}(x) = \Delta^p x_l x_1 \cdots x_{l-1}, \qquad \mathbf{d}^2(x) = \Delta^p x_{l-1} x_l x_1 \cdots x_{l-2}, \qquad \ldots$$

so $\mathbf{d}^l(x) = x$ in this case. If p is odd we get:

$$\mathbf{d}(x) = \Delta^p \tau(x_l) x_1 \cdots x_{l-1}, \qquad \mathbf{d}^2(x) = \Delta^p \tau(x_{l-1}) \tau(x_l) x_1 \cdots x_{l-2}, \qquad \ldots$$

so $\mathbf{d}^{2l}(x) = x$ in this case. We then see that, if x is rigid, iterated cyclings and decyclings correspond to cyclic permutations of the factors in the normal form of x (possibly conjugated by Δ, if p is odd); moreover, when applied to rigid braids, $\mathbf{c}$ and $\mathbf{d}$ are inverses of each other.

2.3. Summit Sets

Let now $x \in \mathbb{B}_n$ be an arbitrary braid (not necessarily rigid). Consider the conjugacy class of x, denoted $x^{\mathbb{B}_n}$, and write $\inf_s(x)$ (resp. $\sup_s(x)$) for the maximal infimum (resp. the minimal supremum) of an element in $x^{\mathbb{B}_n}$. These numbers are known to exist [12], and are called the *summit infimum* and the *summit supremum* of x, respectively. Set $\ell_s(x) = \sup_s(x) - \inf_s(x)$, the *summit length* of x. It is shown in [12] that the elements in $x^{\mathbb{B}_n}$ having the shortest possible normal form are those whose canonical length is precisely $\ell_s(x)$, and they coincide with the elements whose infimum and supremum are equal to $\inf_s(x)$ and $\sup_s(x)$, respectively. The set formed by these elements is called the *supper summit set* of the braid x:

$$SSS(x) = \left\{ y \in x^{\mathbb{B}_n} \mid \ell(y) = \ell_s(x) \right\} = \left\{ y \in x^{\mathbb{B}_n} \mid \inf(y) = \inf_s(x),\ \sup(y) = \sup_s(x) \right\}.$$

Starting from x, it is possible to obtain an element in $SSS(x)$ by applying cyclings and decyclings iteratively. It is known [12] that if $\inf(x) < \inf_s(x)$ then the infimum of x can be increased by iterated cycling. Actually, in this case $\inf(x) < \inf(\mathbf{c}^k(x))$ for some $k < \frac{n(n-1)}{2}$ (see [14]). Hence, every $\frac{n(n-1)}{2}$ iterations either the infimum has increased, or one is sure to have an element whose infimum is the summit infimum.

In the same way, if $\sup(x) > \sup_s(x)$, then the supremum of x can be decreased by iterated decycling [12], and in that case $\sup(x) > \sup(\mathbf{d}^k(x))$ for some $k < \frac{n(n-1)}{2}$ [14]. Hence, every $\frac{n(n-1)}{2}$ iterations either the supremum has decreased, or we are sure to have an element whose supremum is the summit supremum. Since decycling can never decrease the infimum of an element, it follows that starting with any $x \in \mathbb{B}_n$ and applying iterated cycling (until summit infimum is obtained) followed by iterated decycling (until summit supremum is obtained) yields an element $y \in SSS(x)$.

The super summit set $SSS(x)$ is a finite set, but it is usually huge, so smaller subsets of the conjugacy class of x were defined in order to solve the conjugacy problem of x more efficiently. Namely, the *ultra summit set* of x, denoted by $USS(x)$, is a subset of $SSS(x)$ defined as follows [4]:

$$USS(x) = \{y \in SSS(x) \mid \mathbf{c}^m(y) = y \text{ for some } m > 0\}.$$

Since $SSS(x)$ is finite, the subset $USS(x)$ is also finite. It is then clear that one obtains an element is $USS(x)$ by iterated application of cycling, starting from an element in $SSS(x)$, when a repeated element is obtained. Actually, the whole orbit under cycling of an element in $USS(x)$ belongs to $USS(x)$. So $USS(x)$ is a finite set of orbits under cycling.

Notice that every rigid braid belongs to its ultra summit set, as cylings and decyclings are basically cyclic permutations of its factors. It is shown in [15] that, if x is conjugate to a rigid braid and $\ell_s(x) > 1$, then $USS(x)$ coincides with the set of rigid conjugates of x.

There is actually a simpler way, in the general case, to obtain an element in $USS(x)$ starting from x. Instead of using cyclings and decyclings, one can use the following single type of conjugation:

Definition 5 ([5]). *Given $x \in \mathbb{B}_n$, the cyclic sliding of x is defined as $\mathfrak{s}(x) = \mathfrak{p}(x)^{-1} x \, \mathfrak{p}(x)$, where $\mathfrak{p}(x) = \iota(x) \wedge \partial(\varphi(x))$.*

Theorem 1 ([5]). *Given $x \in \mathbb{B}_n$, there are integers $0 \leq k < t$ such that $\mathfrak{s}^k(x) = \mathfrak{s}^t(x)$. For every such pair of integers, one has $\mathfrak{s}^k(x) \in USS(x)$.*

By the above result, one can obtain an element in $USS(x)$ by iterated cyclic sliding starting form x. Furthermore, if x is conjugate to a rigid element (this will be the generic situation, as we will see in Section 2.4), iterated cyclic sliding yields the *shortest* positive conjugating element from x to a rigid element.

Theorem 2 ([5]). *Let $x \in \mathbb{B}_n$ and suppose that x is conjugate to a rigid braid. Then there is an integer $k > 0$ such that $\mathfrak{s}^k(x)$ is rigid. Moreover, the conjugating element α from x to $\mathfrak{s}^k(x)$, that is,*

$$\alpha = \mathfrak{p}(x)\,\mathfrak{p}(\mathfrak{s}(x))\,\mathfrak{p}(\mathfrak{s}^2(x)) \cdots \mathfrak{p}(\mathfrak{s}^{k-1}(x))$$

is the smallest positive element (with respect to $\preccurlyeq$) conjugating x to a rigid element, meaning that for every positive element β such that $\beta^{-1} x \beta$ is rigid, one has $\alpha \preccurlyeq \beta$.

After obtaining one element in $USS(x)$, it is possible to compute all elements in $USS(x)$ together with conjugating elements connecting them. In this way, one solves the conjugacy problem in $\mathbb{B}_n$, as two elements x and y are conjugate if and only if $USS(x) = USS(y)$ or, equivalently, if $USS(x) \cap USS(y) \neq \emptyset$. Then, in order to check whether x and y are conjugate, one can compute the whole set $USS(x)$, and one element $\widetilde{y} \in USS(y)$. Then, x and y are conjugate if and only if $\widetilde{y} \in USS(x)$. By construction, one can even compute a conjugating element from x to y.

In order to understand the forthcoming proofs in this paper, we will need to describe some conjugating elements connecting the elements of $USS(x)$.

Definition 6 ([4]). *Let $x \in \mathbb{B}_n$ and $y \in USS(x)$. A simple non-trivial element $s \in S$ is said to be a **minimal simple element** for y if $y^s \in USS(x)$ and $y^t \notin USS(x)$, for every $1 \prec t \prec s$.*

In [4], Gebhardt showed that for any two elements $y, z \in USS(x)$ there exists a sequence

$$y = y_1 \xrightarrow{c_1} y_2 \xrightarrow{c_2} \cdots \to y_t \xrightarrow{c_t} y_{t+1} = z,$$

where c_i is a minimal simple element for y_i, and $y_{i+1} = c_i^{-1} y_i c_i$, for $i = 1, \ldots, t$. Moreover, he introduced an algorithm to compute all minimal simple elements for a given $y \in USS(x)$. This allows to construct a directed graph Γ_x, whose vertices correspond to elements of $USS(x)$, and whose arrows correspond to minimal simple elements, in such a way that for every minimal simple element s for y, there is an edge with label s from y to $y^s = s^{-1}ys$. By the above discussion, it follows that Γ_x is a connected graph, and this is why $USS(x)$ can be computed starting with a single vertex, iteratively computing the minimal simple elements corresponding to each known vertex, until all vertices are obtained.

We will later see that, generically, ultra summit sets are really small. Actually, they usually have a very simple structure, that we explain now.

Lemma 1 ([16]). *Let $y \in USS(x)$ with $\ell(y) > 0$ and let s be a minimal simple element for y. Then, s is a prefix of either $\iota(y)$ or $\partial(\varphi(y))$, or both.*

The above lemma allows us to classify the arrows in Γ_x into two groups: a directed edge labelled by s starting at $y \in USS(x)$ is black (resp. grey), if s is a prefix of $\iota(x)$ (resp. of $\partial(\varphi(y))$). In principle, an edge could be of both colors at the same time (a bi-colored arrow, whose label is a prefix of both $\iota(x)$ and $\partial(\varphi(x))$), but not in the case of rigid braids, as $\iota(x) \wedge \partial(\varphi(x)) = 1$ if x is rigid. Actually, this is a necessary and sufficient condition:

Lemma 2 ([16]). *A braid $y \in USS(x)$ with $\ell(y) > 0$ is rigid if and only if none of the edges starting at y is bi-colored.*

Definition 7. *Given a braid $x \in \mathbb{B}_n$, its associated $USS(x)$ is **minimal** if $\ell_s(x) > 1$ and, for every vertex y in the graph Γ_x, there are exactly two directed edges starting at y, a black one labeled $\iota(y)$ and a grey one labeled $\partial(\varphi(y))$.*

Notice that, as a consequence of Lemma 2, if $USS(x)$ is minimal then all elements in $USS(x)$ are rigid. Moreover, the arrow labeled $\iota(y)$ corresponds to a cycling of y, and the arrow labeled $\partial(\varphi(y))$ corresponds to a *twisted decycling* of y, meaning a decycling followed by the automorphism τ. This implies that, if $USS(x)$ is minimal, the elements of $USS(x)$ are obtained from y by applying **c** and $\tau \circ \mathbf{d}$ in every possible way. Since y is rigid, cyclings and decyclings basically correspond to cyclic permutations of the factors. Therefore, if $USS(x)$ is minimal, it consists of either two orbits under cycling (conjugate to each other by Δ), or one orbit under cycling (conjugate to itself by Δ). If the infimum of y is even, the orbit of y has at most $\ell(y) = \ell_s(x) \leq \ell(x)$ elements, so the size of $USS(x)$ is at most $2\ell(x)$. If the infimum of y is odd, the orbit of y has at most $2\ell(y) \leq 2\ell(x)$ elements, and it is conjugate to itself by Δ, so it is the only orbit. Therefore, in any case, if $USS(x)$ is minimal it has at most $2\ell(x)$ elements.

Remark 1. *In order to see whether $USS(x)$ is minimal, one should a priori check the condition in Definition 7 for every element in $USS(x)$. But it is actually shown in ([17], Theorem 4.6) that, given $y \in USS(x)$, the set $USS(x)$ is minimal if and only if $\ell(y) > 1$ and the minimal simple elements for y are precisely $\iota(y)$ and $\partial(\varphi(y))$. Hence, one just needs to compute the minimal elements for a single arbitrary element $y \in USS(x)$.*

Let us see that this case, in which $USS(x)$ is so small and has such a simple structure, is generic.

2.4. Generic Braids

Since $\mathbb{B}_n$ is an infinite set, it is necessary to explain what we mean by 'picking a random braid' or by saying that a braid is 'generic'. Even if we fix the subset of braids of a given length, we must specify if we choose braids from the subset with a uniform distribution, or if we pick braids by choosing a random walk in the Cayley graph, which are the two usual situations.

We will consider the Cayley graph of the braid group $\mathbb{B}_n$, taking as generators the simple braids, and assume that each edge of the Cayley graph has length 1, so it becomes a metric space. Let us point out that left normal forms of braids are closely related to geodesics in this Cayley graph [18].

Now let $B(r)$ denote the ball of radius r centered at the trivial braid 1. As the number of simple braids is finite, the set $B(r)$ is a finite subset of $\mathbb{B}_n$. We will consider the uniform distribution within this set. It turns out that 'most' elements in $B(r)$ have a very simple ultra summit set:

Theorem 3 ([17]). *The proportion of braids in $B(r)$ whose ultra summit set is minimal tends to 1 exponentially fast, as r tends to infinity.*

This is why we can say that the ultra summit set of a 'generic braid' is minimal. Moreover, the above result was obtained by refining the following theorem, which gives some important information concerning the elements in $B(r)$. We have simplified the statement to adapt it to our situation:

Theorem 4 ([19]). *The proportion of braids x in $B(r)$ which are conjugate to a rigid braid $y = \alpha^{-1}x\alpha$, in such a way that α is a positive braid with $\ell(\alpha) < \ell(x)$, tends to 1 exponentially fast, as r tends to infinity.*

Therefore, not only generic braids have minimal ultra summit sets (made of rigid braids), but one can also obtain a rigid conjugate of a generic braid x very fast, applying iterated cyclic sliding to x. By Theorem 2, the obtained conjugating element will be the smallest possible positive conjugator, so its canonical length will be smaller than $\ell(x)$. Once that a rigid conjugate y (which belongs to $USS(x)$) is obtained, one can compute the whole $USS(x)$ very fast, as it consists of at most $2\ell(x)$ elements, connected by cyclings and twisted decyclings. This is why solving the conjugacy problem in braid groups is generically very fast.

We will also be interested in the centralizer $Z(x)$ of a braid x. Notice that if $y = \alpha^{-1}x\alpha$, then $Z(y) = \alpha^{-1}Z(x)\alpha$. Therefore, knowing $Z(y)$ is equivalent to knowing $Z(x)$, via α. We will then be interested in $Z(y)$ for $y \in USS(x)$.

Definition 8. *Let $x \in \mathbb{B}_n$ and $y \in USS(x)$, and let t be the smallest positive integer such that $\mathbf{c}^t(y) = y$. Denote $p_i := \iota(\mathbf{c}^{i-1}(y))$ the positive element conjugating $\mathbf{c}^{i-1}(y)$ to $\mathbf{c}^i(y)$, for $i = 1, \ldots, t$. Then the preferred cycling conjugator of y is defined as*
$$PC(y) = p_1 p_2 \cdots p_t.$$
In other words, $PC(y)$ corresponds to the conjugating element along the whole cycling orbit of y. By construction, $PC(y)$ commutes with y.

In the generic case (when $USS(x)$ is minimal), it turns out that $Z(x)$ is isomorphic to $\mathbb{Z}^2$, and one can describe the generators of $Z(y)$ for any $y \in USS(x)$ (and thus of $Z(x)$) in a very explicit way:

Theorem 5 ([17]). *Let $x \in \mathbb{B}_n$ and $y \in USS(x)$. Let $PC(y) = p_1 \cdots p_t$ as above. If $USS(x)$ is minimal, then all elements in $USS(x)$ are rigid, $Z(x) \simeq Z(y) \simeq \mathbb{Z}^2$, and one of the following conditions holds:*

(i) $USS(x)$ *has two orbits under cycling, conjugate to each other by Δ, and $Z(y) = \langle \Delta^2, PC(y) \rangle$.*

(ii) $USS(x)$ *has one orbit under cycling, conjugate to itself by Δ, and:*

- *If $\tau(y) = y$, then $Z(y) = \langle \Delta, PC(y) \rangle$.*
- *If $\tau(y) \neq y$, then t is even and $Z(y) = \langle \Delta^2, p_1 \cdots p_{\frac{t}{2}} \Delta^{-1} \rangle$.*

3. k-th Root Problem

Now we come to the central problem in this paper: given $x \in \mathbb{B}_n$ and an integer $k > 1$, find a k-th root of x. In other words, we want to either find $a \in \mathbb{B}_n$ such that $a^k = x$, or show that such a braid does not exist.

Notice that if $a^k = x$ then a belongs to $Z(x)$, the centralizer of x. It is interesting to know that finding a single solution a to the k-th root equation is basically the same as finding all possible solutions, as the complete set of solutions coincides with the conjugacy class of a in $Z(x)$:

Proposition 1. *Let $a, x \in \mathbb{B}_n$ be such that $a^k = x$ for some integer $k > 1$. Then the set $\sqrt[k]{x}$ of k-th roots of x is precisely*
$$\sqrt[k]{x} = a^{Z(x)} = \left\{ b \in \mathbb{B}_n \mid b = u^{-1}au, \; u \in Z(x) \right\}.$$

Proof. In [20], the second author proved that the k-th root of a braid is unique, up to conjugacy. That is, if $a, b \in \mathbb{B}_n$ satisfy $a^k = b^k = x$, then $a = u^{-1}bu$ for some $u \in \mathbb{B}_n$. Then one has $x = b^k = u^{-1}a^k u = u^{-1}xu$, and hence $u \in Z(x)$. This proves that $\sqrt[k]{x} \subset a^{Z(x)}$.

On the other hand, if $b = a^{Z(x)}$ and we write $b = u^{-1}au$ for some $u \in Z(x)$, we have $b^k = u^{-1}a^k u = u^{-1}xu = x$, so $b \in \sqrt[k]{x}$. □

Observe that $a^k = x$ if and only if $(\alpha^{-1}a\alpha)^k = \alpha^{-1}x\alpha$ for any $\alpha \in \mathbb{B}_n$. Hence, given x, it suffices to solve the k-th root problem for any conjugate of x, for instance for some $y \in USS(x)$.

We will focus our attention in the generic case in which $USS(x)$ is minimal. Recall from Theorem 5 that in this case $Z(x) \simeq Z(y) \simeq \mathbb{Z}^2$. If we express the centralizer of y as $Z(y) = \langle v, w \rangle$, where v and w commute, we know that y has the form $y = v^c w^d$, for some $c, d \in \mathbb{Z}$ (and that this expression is unique, as any other expression would yield a different element of $Z(y)$). If we are able to express y in this way, then the k-th root problem is trivially solved:

Proposition 2. *Let $x \in \mathbb{B}_n$. Let $y \in USS(x)$ and suppose that $USS(x)$ is minimal. Let $Z(y) = \langle v, w \rangle$ and let $c, d \in \mathbb{Z}$ be such that $y = v^c w^d$. Then y admits a k-th root if and only if both c and d are multiples of k, and in this case the only k-th root of y is:*
$$a = v^{\frac{c}{k}} w^{\frac{d}{k}}.$$

Proof. We know from Theorem 5 that $Z(y) \simeq \mathbb{Z}^2$, so it is abelian. Hence, by Proposition 1, if a k-th root a of y exists then $\sqrt[k]{y} = a^{Z(y)} = \{a\}$. Therefore, if a k-th root exists, it is unique.

Suppose that the k-th root problem for y has a solution $a \in \mathbb{B}_n$. Then $a \in Z(y)$, and hence $a = v^r w^s$ for some $r, s \in \mathbb{Z}$. But since v and w commute, we have:

$$v^c w^d = y = a^k = (v^r w^s)^k = v^{rk} w^{sk}.$$

This implies that c and d are multiples of k, and that $a = v^r w^s = v^{\frac{c}{k}} w^{\frac{d}{k}}$.

Conversely, if c and d are multiples of k, we write $c = rk$ and $d = sk$ for some integers r, s, and we consider the element $a = v^r w^s$. Since v and w commute, it follows that $a^k = y$. □

By the above result, it follows that the only difficulty in solving the k-th root problem, in the generic case in which $USS(x)$ is minimal, is to express some $y \in USS(x)$ in terms of the generators of $Z(y)$. We know from Theorem 5 that there are three possible cases, depending on whether $USS(x)$ has two orbits under cycling, or has one orbit with $\tau(y) = y$, or has one orbit with $\tau(y) \neq y$. The three following results address each case:

Proposition 3. *Let $x \in \mathbb{B}_n$, and let $y = \Delta^p y_1 \cdots y_l \in USS(x)$, written in left normal form. Suppose that $USS(x)$ is minimal. Suppose also that $USS(x)$ has two orbits under cycling, conjugate to each other by Δ. Let $v = \Delta^2$ and $w = PC(y) = p_1 \cdots p_t$, so:*

$$Z(y) = \langle v, w \rangle = \langle \Delta^2, PC(y) \rangle.$$

If we write $c = p/2$ and $d = l/t$, then c and d are integers and we have: $y = v^c w^d$.

Proof. We know that, since $USS(x)$ is minimal, it consists of rigid elements. Hence iterated cycling corresponds to a cyclic permutation of the factors in the normal form of y (with possible conjugations by Δ, if p is odd).

Suppose that p is odd. Then $\mathbf{c}^l(y)$ is obtained from y by cyclically permuting all its l factors, conjugating all of them by Δ. Hence $\mathbf{c}^l(y) = \tau(y)$. This implies that $\tau(y) = \Delta^{-1} y \Delta$ is in the same orbit of y under cycling, but this is a contradiction with the hypotheses, as $USS(x)$ has two distinct orbits (the one containing y and the one containing $\tau(y)$). Therefore p is even.

Since p is even, iterated cyclings of y correspond exactly to cyclic permutations of the factors of y. By definition, t is the smallest positive integer such that $\mathbf{c}^t(y) = y$, and it is then clear that $\mathbf{c}^m(y) = y$ for some positive integer m if and only if m is a multiple of t. Since $\mathbf{c}^l(y) = y$, we finally obtain that l is a multiple of t. Then the normal form of y is as follows:

$$y = \Delta^p y_1 \cdots y_l = \Delta^p (y_1 \cdots y_t)(y_1 \cdots y_t) \cdots (y_1 \cdots y_t),$$

where $PC(y) = y_1 \cdots y_t$, and there are l/t parenthesized factors.

Now, if we write $c = p/2$ and $d = l/t$, these numbers are integers and we have:

$$v^c w^d = (\Delta^2)^c (PC(y))^d = \Delta^{2c} (y_1 \cdots y_t)^d = \Delta^p y_1 \cdots y_l = y.$$

□

Proposition 4. *Let $x \in \mathbb{B}_n$, and let $y = \Delta^p y_1 \cdots y_l \in USS(x)$, written in left normal form. Suppose that $USS(x)$ is minimal. Suppose also that $USS(x)$ has one orbit under cycling, conjugate to itself by Δ, and that $\tau(y) = y$. Let $v = \Delta$ and $w = PC(y) = p_1 \cdots p_t$, so:*

$$Z(y) = \langle v, w \rangle = \langle \Delta, PC(y) \rangle.$$

If we write $c = p$ and $d = l/t$, then c and d are integers and we have: $y = v^c w^d$.

Proof. We know that the left normal form of $\tau(y)$ is $\Delta^p \tau(y_1) \cdots \tau(y_l)$. Since $\tau(y) = y$, the normal forms of y and $\tau(y)$ must coincide, hence $\tau(y_i) = y_i$ for $i = 1, \ldots, l$.

This implies that iterated cyclings correspond to cyclic permutations of the factors of y. We do not care about the parity of p, as every factor of y is invariant under τ. It then follows that $PC(y) = y_1 \cdots y_t$, that t divides l and that the normal form of y is:

$$y = \Delta^p y_1 \cdots y_l = \Delta^p (y_1 \cdots y_t)(y_1 \cdots y_t) \cdots (y_1 \cdots y_t),$$

where there are l/t parenthesized factors.

Now, if we write $c = p$ and $d = l/t$, these numbers are integers and we have:

$$v^c w^d = \Delta^c (PC(y))^d = \Delta^c (y_1 \cdots y_t)^d = \Delta^p y_1 \cdots y_l = y.$$

□

Proposition 5. *Let $x \in \mathbb{B}_n$, and let $y = \Delta^p y_1 \cdots y_l \in USS(x)$, written in left normal form. Suppose that $USS(x)$ is minimal. Suppose also that $USS(x)$ has one orbit under cycling, conjugate to itself by Δ, and that $\tau(y) \ne y$. Let $v = \Delta^2$, $PC(y) = p_1 \cdots p_t$ and $w = p_1 \cdots p_{\frac{t}{2}} \Delta^{-1}$ (recall from Theorem 5 that t is even), so:*

$$Z(y) = \langle v, w \rangle = \langle \Delta, p_1 \cdots p_{\frac{t}{2}} \Delta^{-1} \rangle.$$

If we write $c = \frac{pt+2l}{2t}$ and $d = \frac{2l}{t}$, then c and d are integers and we have: $y = v^c w^d$.

Proof. We know from Theorem 5 that t is even, but let us see why this holds. We know that there exists some $m > 0$ so that $\tau(y) = \mathbf{c}^m(y)$; we take m as small as possible, and this implies that $\mathbf{c}^r(y) \ne y$ for $0 < r < m$. Now, it follows from their own definitions that τ and $\mathbf{c}$ commute, and therefore $y = \tau^2(y) = \tau(\mathbf{c}^m(y)) = \mathbf{c}^m(\tau(y)) = \mathbf{c}^{2m}(y)$. This implies that the length of the cycling orbit of y is a divisor of $2m$. It cannot be m (as $\mathbf{c}^m(y) = \tau(y) \ne y$), and it cannot be smaller than m (as $\mathbf{c}^r(y) \ne y$ for every $r < m$). Therefore, the length of the orbit is precisely $t = 2m$. The generators of $Z(y)$ are then $v = \Delta^2$ and $w = p_1 \cdots p_m \Delta^{-1}$.

We consider now two cases, depending on the parity of p. If p is even, since the first m cyclings transform y into $\tau(y)$, it follows that the left normal form of y is:

$$y = \Delta^p (y_1 \cdots y_m)(\tau(y_1) \cdots \tau(y_m)) \cdots (y_1 \cdots y_m)(\tau(y_1) \cdots \tau(y_m)).$$

Then $l = 2rm$ for some positive integer r.

Recall that $PC(y)$ is the product of the first $t = 2m$ conjugating elements for cycling. The first m conjugating elements are $y_1, \ldots, y_m$, so $p_i = y_i$ for $i = 1, \ldots, m$. The following m conjugating elements are $\tau(y_1), \ldots, \tau(y_m)$. Hence, we have that

$$\begin{aligned}
PC(y) &= p_1 \cdots p_t \\
&= y_1 \cdots y_m \tau(y_1) \cdots \tau(y_m) \\
&= y_1 \cdots y_m \ \tau(y_1 \cdots y_m) \\
&= p_1 \cdots p_m \Delta^{-1} p_1 \cdots p_m \Delta \\
&= \left(p_1 \cdots p_m \Delta^{-1}\right) \left(p_1 \cdots p_m \Delta^{-1}\right) \Delta^2 \\
&= w^2 v.
\end{aligned}$$

Therefore, if p is even:

$$y = \Delta^p PC(y)^r = v^{\frac{p}{2}} \left(w^2 v\right)^r = v^{\frac{p}{2}+r} w^{2r} = v^c w^d,$$

where $c = \frac{pt+2l}{2t}$ and $d = \frac{2l}{t}$ (recall that $l = 2rm = rt$).

Consider now the case when p is odd. In this case, the left normal form of y is:

$$y = \Delta^p \ (y_1 \cdots y_m) \ (\tau(y_1) \cdots \tau(y_m)) \ \cdots \ (y_1 \cdots y_m) \ (\tau(y_1) \cdots \tau(y_m)) \ (y_1 \cdots y_m).$$

Then $l = (2r+1)m$ for some positive integer r.

As before, $PC(y)$ is the product of the first $t = 2m$ conjugating elements for cycling, but this time the first m conjugating elements for cycling are $\tau(y_1), \ldots, \tau(y_m)$, and therefore $p_i = \tau(y_i)$ for $i = 1, \ldots, m$. The following m conjugating elements are $y_1, \ldots, y_m$, so we have:

$$\begin{aligned}
PC(y) &= p_1 \cdots p_t \\
&= \tau(y_1) \cdots \tau(y_m) y_1 \cdots y_m \\
&= p_1 \cdots p_m \ \tau(p_1 \cdots p_m) \\
&= p_1 \cdots p_m \Delta^{-1} p_1 \cdots p_m \Delta \\
&= \left(p_1 \cdots p_m \Delta^{-1}\right) \left(p_1 \cdots p_m \Delta^{-1}\right) \Delta^2 \\
&= w^2 v.
\end{aligned}$$

Hence $PC(y) = w^2 v$ also when p is odd. Finally, we have:

$$\begin{aligned}
y &= \Delta^p (y_1 \cdots y_m)(\tau(y_1) \cdots \tau(y_m)) \cdots (y_1 \cdots y_m)(\tau(y_1) \cdots \tau(y_m))(y_1 \cdots y_m) \\
&= (\tau(y_1) \cdots \tau(y_m))(y_1 \cdots y_m) \cdots (\tau(y_1) \cdots \tau(y_m))(y_1 \cdots y_m)\Delta^p(y_1 \cdots y_m) \\
&= PC(y)^r \Delta^p (y_1 \cdots y_m) \\
&= PC(y)^r \Delta^{p+1} \Delta^{-1}(y_1 \cdots y_m) \\
&= PC(y)^r \Delta^{p+1} (p_1 \cdots p_m)\Delta^{-1} \\
&= (w^2 v)^r v^{\frac{p+1}{2}} w \\
&= v^{\frac{2r+1+p}{2}} w^{2r+1} \\
&= v^c w^d,
\end{aligned}$$

105

where $c = \frac{pt+2l}{2t}$ and $d = \frac{2l}{t}$ (recall that $2l = 2(2r+1)m = (2r+1)t$ in this case). □

4. An Algorithm to Find the k-th Root of a Braid

We end this paper by providing a detailed algorithm that summarizes the results from the previous section, together with a study of its complexity.

The results of the previous section are valid when $USS(x)$ is minimal (which is the generic case). In order to have an algorithm which always succeeds in finding the k-th root of a braid x, we need to include instructions on what to do if $USS(x)$ is not minimal. In those cases, one can use the algorithm in [7], which finds the k-th root of x in any case, considering the Garside group $G = \mathbb{Z} \ltimes (\mathbb{B}_n)^k$, where $\mathbb{Z} = \langle \delta \rangle$ acts on $(\mathbb{B}_n)^k$ by cyclic permutation of the coordinates. S. J. Lee shows that the braid x has a k-th root if and only if the ultra summit set of $\delta(x, 1, \ldots, 1)$ in G has an element of the form $\delta(h, \ldots, h)$. Hence, computing an ultra summit set in such a group also solves the root extraction problem in $\mathbb{B}_n$. It is not clear to us how big these ultra summit sets are in generic cases, while the algorithm presented in this paper is very simple, and generically very fast.

If one is not interested in programming the algorithm in [7], one could tell our algorithm to return 'fail' when $USS(x)$ is not minimal, obtaining an algorithm which will succeed only in the generic case. In any case, we present now the main result:

Theorem 6. *There is an algorithm that takes as input a braid $x = \Delta^p x_1 \ldots, x_l \in \mathbb{B}_n$ written in left normal form, and a positive integer $k > 1$, and finds a braid $a \in \mathbb{B}_n$ such that $a^k = x$, or guarantees that such a braid does not exist, whose generic-case complexity is $O(l(l+n)n^3 \log n)$.*

Proof. Algorithm 1, which uses the results from the previous section, constitutes a proof of the theorem. Let us describe it in detail.

The input is a braid $x = \Delta^p x_1 \cdots x_l \in \mathbb{B}_n$ in left normal form and an integer $k > 1$. First (lines 2–5), the algorithm applies iterated cyclic sliding to x, checking at each iteration whether the resulting braid y is rigid. As we will now see, if the algorithm applies cyclic sliding $l\left(\frac{n(n-1)}{2} - 1\right)$ times and no rigid braid is obtained, then we are not in the generic case stated in Theorem 4, hence the algorithm in [7] is applied. The number $l\left(\frac{n(n-1)}{2} - 1\right)$ is precisely l times the length of Δ minus one. Recall from Theorem 4 that in the generic case there is a positive element α conjugating x to a rigid braid, such that $\ell(\alpha) < \ell(x) = l$. If α is the smallest possible one, there is no Λ in its normal form. Hence, the length of α in terms of atoms (σ_i's) is at most $l\left(\frac{n(n-1)}{2} - 1\right)$. Now, from Theorem 2 we know that the smallest positive conjugator to a rigid braid is obtained by iterated cyclic sliding. Since at every iteration the conjugating element gets bigger, if we are in the generic case we must obtain a rigid element in at most $l\left(\frac{n(n-1)}{2} - 1\right)$ iterations, as we claimed.

If the braid y obtained after the loop in lines 2-5 is rigid, as the algorithms stores the conjugating elements for cyclic sliding at each iteration, we will have a braid α such that $\alpha^{-1} x \alpha = y$.

Now the algorithm checks whether $USS(y)$ is minimal (the generic case we are interested in), as explained in Remark 1, checking whether the minimal simple elements for y are precisely $\iota(y)$ and $\partial(\varphi(y))$.

In general, it is not known how fast it is to compute the minimal simple elements for a given arbitrary braid y. But if y is rigid, one can easily find the minimal simple elements for y. We know that every such element must be a prefix of either $\iota(y)$ or $\partial(\varphi(y))$. For every generator σ_i, one can consider $\sigma_i^{-1} y \sigma_i$ and apply iterated cyclic sliding to it, until it becomes rigid. The obtained conjugating element is the smallest conjugating element from y to a rigid braid, having σ_i as a prefix. We do this for all σ_i which are prefixes of $\iota(y)$, and either we find a conjugating element which is a proper prefix of $\iota(y)$ (in which case $\iota(y)$ is not minimal), or we have shown that $\iota(y)$ is minimal. Then we do the same for all generators which are prefixes of $\partial(\varphi(y))$. The number of iterations in each case is bounded by the length of $\iota(y)$ (resp. $\partial(\varphi(y))$), which are simple elements, while the total number of generators is $n-1$. So the total number of cyclic slidings used to check whether $\iota(y)$ and $\partial(\varphi(y))$ are minimal (and hence whether $USS(y)$ is minimal) is $O(n^3)$.

If $USS(y)$ is not minimal, we are not in the generic case stated in Theorem 4, hence the algorithm in [7] is applied. Otherwise, we are in one of the situations described in Propositions 3–5. The rest of the algorithm just applies these propositions together with Proposition 2: after decomposing y in the form $y = v^c w^d$, it checks whether both c and d are multiples of k. If this is the case, then $v^{\frac{c}{k}} w^{\frac{d}{k}}$ is the (unique) k-th root of y, and since $x = \alpha y \alpha^{-1}$, it follows that $\alpha v^{\frac{c}{k}} w^{\frac{d}{k}} \alpha^{-1}$ is the desired k-th root of x; otherwise, the algorithm returns the sentence "A k-th root does not exist".

We study now the complexity of our algorithm, assuming that we are in the generic case in which $USS(x)$ is minimal, and we can quickly conjugate x to a rigid braid. Computing the complement or applying τ to a simple element is $O(n)$, and computing $s \wedge t$ for two simple elements s and t is $O(n \log n)$ ([13], Proposition 9.5.1). Starting with an element y in left normal form, computing $\mathfrak{s}(y)$ consists of computing a complement ($\partial(\varphi(y))$), a meet ($\iota(x) \wedge \partial(\varphi(x))$) and the normal form of the conjugate of y by a simple element of length at most l (which is $O(ln \log n)$). Hence the total complexity of applying a cyclic sliding is $O(ln \log n)$.

The first loop (lines 2–5) is repeated $O(ln^2)$ times, checking the condition takes $O(n \log n)$ and the body of the loop takes $O(ln \log n)$. Hence the total complexity of the loop in lines 2–5 is $O(l^2 n^3 \log n)$.

The "If" statement in lines 6–7 is negligible compared with the previous "while" loop.

Next, in lines 8–9 the algorithm checks whether $\iota(y)$ and $\partial(\varphi(y))$ are minimal, for the rigid element y. By the arguments above, this applies $O(n^3)$ cyclic slidings, hence the total complexity of this step is $O(ln^4 \log n)$.

In line 11 and in the loop in lines 12–15, some cyclings are applied. Since the involved braids are rigid of canonical length at most l, and cycling is just a cyclic permutation of the factors with a possible application of τ to a simple element, this final part of the algorithm is negligible with respect to the previous one.

Therefore, the generic-case complexity of Algorithm 1 is $O(l(l+n)n^3 \log n)$. □

Remark 2. *Although the integers p and k are part of the input, the computed complexity does not involve them, as treating with these integers is usually negligible, in reasonable examples, with respect to the calculated complexity. If p is really big, one should take into account the number $\log p$. The case of k is somehow different, as one would have a positive answer only if k is a divisor of the integers c and d (with $d \neq 0$), which are $O(p+l)$, so it makes no sense to ask for a k-th root of x, in the generic case, if k is too big compared with p and l.*

Algorithm 1: Find a k-th root of a braid x.

Input   : A braid $x \in \mathbb{B}_n$ given in left normal form, and an integer $k > 1$.
Output: A braid $a \in \mathbb{B}_n$ such that $a^k = x$, or the message "*A k-th root does not exist.*".

1   $y := x;\quad l = \ell(x);\quad \alpha = 1 \in \mathbb{B}_n;\quad r = 0 \in \mathbb{Z}$.
2   **while** $\iota(y) \wedge \partial(\varphi(y)) \neq 1$ **and** $r < l\left(\frac{n(n-1)}{2} - 1\right)$ **do**
3 | $\alpha := \alpha\, \mathfrak{p}(y)$;
4 | $y := \mathfrak{s}(y)$;
5 | $r := r + 1$;
6   **if** $\iota(y) \wedge \partial(\varphi(y)) \neq 1$ **then**
7 | y is not rigid. Apply the algorithm in [7];
8   **else if** {*Minimal simple elements for* y} $\neq \{\iota(y), \partial(\varphi(y))\}$ **then**
9 | $USS(y)$ is not minimal. Apply the algorithm in [7];
10 **else**
11 | $y' := \tau(y);\quad z := \mathfrak{c}(y);\quad PC := \iota(y) \in \mathbb{B}_n;\quad t := 1 \in \mathbb{Z};\quad p := \inf(y);\quad l := \ell(y);\quad \text{selfConjugateOrbit} := 0$;
12 | **while** $z \neq y$ **and** $z \neq y'$ **do**
13 | $PC := PC\, \iota(z)$;
14 | $z := \mathfrak{c}(z)$;
15 | $t := t + 1$;
16 | **if** $z = y'$ **then**
17 | $\text{selfConjugateOrbit} := 1$;
18 | **if** $\text{selfConjugateOrbit} = 0$ **then**
19 | $c := p/2$;
20 | $d := l/t$;
21 | **if** $k|c$ **and** $k|d$ **then**
22 | $v := \Delta^2$;
23 | $w := PC$;
24 | **return** $\alpha v^{\frac{c}{k}} w^{\frac{d}{k}} \alpha^{-1}$;
25 | **else**
26 | **return** "*A k-th root does not exist.*";
27 | **else if** $\text{selfConjugateOrbit} = 1$ **and** $y = y'$ **then**
28 | $c := p$;
29 | $d := l/t$;
30 | **if** $k|c$ **and** $k|d$ **then**
31 | $v := \Delta$;
32 | $w := PC$;
33 | **return** $\alpha v^{\frac{c}{k}} w^{\frac{d}{k}} \alpha^{-1}$;
34 | **else**
35 | **return** "*A k-th root does not exist.*";
36 | **else if** $\text{selfConjugateOrbit} = 1$ **and** $y \neq y'$ **then**
37 | $t := 2t$;
38 | $c := \frac{pt + 2l}{2t}$;
39 | $d := \frac{2l}{t}$;
40 | **if** $k|c$ **and** $k|d$ **then**
41 | $v := \Delta$;
42 | $w := PC\, \Delta^{-1}$;
43 | **return** $\alpha v^{\frac{c}{k}} w^{\frac{d}{k}} \alpha^{-1}$;
44 | **else**
45 | **return** "*A k-th root does not exist.*";

Author Contributions: Investigation and writing: M.C., J.G.-M. and M.S.

Funding: Authors partially supported by the Spanish research project MTM2016-76453-C2-1-P and FEDER. First author was also supported by EPSRC New Investigator Award EP/S010963/1. Third author was also supported by the Basque Government grant IT974-16 and Centro de Estudios Avanzados en Física, Matemáticas y Computación de la Universidad de Huelva.

Conflicts of Interest: The authors declare no conflict of interest.

References

1. Dehornoy, P. Braid-based cryptography. In *Group Theory, Statistics, and Cryptography*; Volume 360 of Contemporary Mathematics; American Mathematical Society: Providence, RI, USA, 2004; pp. 5–33.
2. Anshel, I.; Anshel, M.; Goldfeld, D. An algebraic method for public-key cryptography. *Math. Res. Lett.* **1999**, *6*, 287–291. [CrossRef]
3. Ko, K.H.; Lee, S.J.; Cheon, J.H.; Han, J.W.; Kang, J.; Park, C. New public-key cryptosystem using braid groups. In *Advances in Cryptology — CRYPTO 2000*; Bellare, M., Ed.; Springer: Berlin/Heidelberg, Germany, 2000; pp. 166–183.
4. Gebhardt, V. A new approach to the conjugacy problem in Garside groups. *J. Algebra* **2005**, *292*, 282–302. [CrossRef]
5. Gebhardt, V.; González-Meneses, J. Solving the conjugacy problem in Garside groups by cyclic sliding. *J. Symb. Comput.* **2010**, *45*, 629–656. [CrossRef]
6. Gebhardt, V.; González-Meneses, J. The cyclic sliding operation in Garside groups. *Math. Z.* **2010**, *265*, 85–114. [CrossRef]
7. Lee, S.-J. Garside groups are strongly translation discrete. *J. Algebra* **2007**, *309*, 594–609. [CrossRef]
8. Sibert, H. Extraction of roots in Garside groups. *Comm. Algebra* **2002**, *30*, 2915–2927. [CrossRef]
9. Dehornoy, P. *Foundations of Garside Theory*; Volume 22 of *EMS Tracts in Mathematics*; Digne, F., Godelle, E., Krammer, D., Michel, J., Eds.; European Mathematical Society (EMS): Zürich, Switzerland, 2015.
10. Artin, E. Theory of Braids. *Ann. Math.* **1947**, *48*, 101–126. [CrossRef]
11. Chow, W.-L. On the algebraical braid group. *Ann. Math.* **1948**, *49*, 654–658. [CrossRef]
12. Elrifai, E.A.; Morton, H.R. Algorithms for positive braids. *Q. J. Math.* **1994**, *45*, 479–497. [CrossRef]
13. Epstein, D.A.; Cannon, J.W.; Holt, D.F.; Levy, S.V.; Paterson, M.S.; Thurston, W.P. *Word Processing in Groups*; A. K. Peters, Ltd.: Natick, MA, USA, 1992.
14. Birman, J.S.; Ko, K.H.; Lee, S.J. The Infimum, Supremum, and Geodesic Length of a Braid Conjugacy Class. *Adv. Math.* **2001**, *164*, 41–56. [CrossRef]
15. Birman, J.S.; Gebhardt, V.; González-Meneses, J. Conjugacy in Garside groups. I. Cyclings, powers and rigidity. *Groups Geom. Dyn.* **2007**, *1*, 221–279. [CrossRef]
16. Birman, J.S.; Gebhardt, V.; González-Meneses, J. Conjugacy in Garside groups II: Structure of the ultra summit set. *Groups Geom. Dyn.* **2008**, *2*, 13–61. [CrossRef]
17. González-Meneses, J.; Valladares, D. On the centralizer of generic braids. *J. Group Theory* **2018**, *21*, 973–1000. [CrossRef]
18. Charney, R. Artin groups of finite type are biautomatic. *Math. Ann.* **1992**, *292*, 671–683. [CrossRef]
19. Caruso, S.; Wiest, B. On the genericity of pseudo-Anosov braids II: Conjugations to rigid braids. *Groups Geom. Dyn.* **2017**, *11*, 549–565. [CrossRef]
20. González-Meneses, J. The n-th root of a braid is unique up to conjugacy. *Algebraic Geom. Topol.* **2003**, *3*, 1103–1118. [CrossRef]

© 2019 by the authors. Licensee MDPI, Basel, Switzerland. This article is an open access article distributed under the terms and conditions of the Creative Commons Attribution (CC BY) license (http://creativecommons.org/licenses/by/4.0/).

Article

A Computational Approach to Verbal Width for Engel Words in Alternating Groups [†]

Jorge Martínez Carracedo

School of Computing, Jordanstown Campus, Ulster University, Northern Ireland BT37 0QB, UK; j.martinez-carracedo@ulster.ac.uk

† This paper is an extended version of our paper published in Lecture Notes of the XVII 'Jacques-Louis Lions' Spanish-French School. Computational Mathematics, Numerical Analysis and Applications (Springer, 2017).

Received: 18 June 2019; Accepted: 2 July 2019; Published: 3 July 2019

Abstract: It is known that every element in the alternating group A_n, with $n \geq 5$, can be written as a product of at most two Engel words of arbitrary length. However, it is still unknown if every element in an alternating group is an Engel word of Arbitrary length. In this paper, a different approach to this problem is presented, getting new results for small alternating groups.

Keywords: group theory; symmetry; Engel words; alternating group

1. Introduction

In recent times, many novel cryptosystems based on Group Theory have been proposed. Even when the ideas behind these group-based cryptosystems are interesting in their own right, these cryptosystems cannot yet compete with more standardized schemes such as Diffie-Hellman or RSA.

The word problem and the conjugacy problem are two of the fundamental decision problems in group theory proposed by Max Dehn in 1911 [1]. The study and understanding of these problems in particular groups have played an important role on group-based cryptosystems.

Braid groups (see Reference [2]), for example, are the mathematical structures behind many cryptographic schemes proposed in the last thirty years. One of the main reasons why these groups are suitable to be used in cryptography is the existence of *normal forms* that facilitates an efficient solution of the word problem [3].

The conjugacy search problem (i.e., given two elements x and y of a group G that are conjugated, find the element $z \in G$ such that $x = y^z$) is the ground from which it is possible to build an scheme similar to ElGalmal in braid groups ([4]).

Solving the conjugacy problem in braid groups is the most direct way to attack this scheme. Garside ([5]) proposed the first algorithm in 1969 to solve this problem in a braid group. However, Garside's proposal is not efficient and a polynomial time algorithm has not been found yet. Heuristics algorithms (as proposed by Hofheinz and Steinwandt in Reference [6]) have achieved a large quota of success though.

Another example of key agreement protocol where the conjugacy search problem plays an important role was proposed by Anshel et al. [7] in 1999. Broadly speaking, in this protocol the two parties agree on a common key by computing a commutator. It was first proposed for braid groups for two reasons: the existence of normal forms and the fact that the conjugacy search problem is considered difficult in these groups.

Cryptography based on group theory has brought about new and interesting pure mathematical questions. The word problem and the conjugacy problem play an important role in some cryptographic schemes based in group theory. Therefore, its study in particular groups seems unavoidable.

Let us consider now an arbitrary group G and a word in the free group of rank r, $\omega \in \mathbb{F}_r$, with r a natural number. We can define the map

$$\omega : \overbrace{G \times \cdots \times G}^{r} \longrightarrow G$$

where each tuple $(g_1, g_2, ..., g_r)$ is mapped to $\omega(g_1, g_2, ..., g_r)$.

We denote the image of the map ω by $\omega(G)$. The verbal subgroup of G related to ω is defined as the subgroup generated by $\omega(G)$.

The surjectivity of the map ω, the cardinality of the set $\omega(G)$, if the verbal subgroup $\langle \omega(G) \rangle = G$ or if it is possible to find a constant k for which $\omega(G)^k = \langle \omega(G) \rangle$ are essential questions to answer.

In 1951, O. Ore proved ([8]) that every element in an alternating group A_n, with $n \geq 5$, can be written as a commutator in A_n.

In other words, Ore's result states that if we consider the word $\tau := x_1^{-1} x_2^{-1} x_1 x_2$ in the free group of rank 2, $\mathbb{F}_2$, then $\tau(A_n) = A_n$, for every $n \geq 5$.

In the same work, he proposed a conjecture: "Every element in a finite simple group G is a commutator in G". This is known as the Abstract Ore's Conjecture and was an open question until 2010.

One initial progress on this conjecture was done in 1994 by Wilson [9]. He proved that for any finite simple group, there exists a constant k such that, $\tau(G)^k = G$.

In this line of work, some new results were obtained considering the word $\xi := x^n$, with n a natural number. In 1996, Martínez and Zelmanov [10] and in 1997, Saxl and Wilson [11] proved, independently, that for every finite simple group big enough, there exists a constant k such that $\xi(G)^k = G$.

In 2010, M. W. Liebeck, E. A. O'Brien, A. Shalev and P. H. Tiep [12] published the proof of the Ore Conjecture. They proved that for every finite simple group G, $G = \tau(G)$, where $\tau := x_1^{-1} x_2^{-1} x_1 x_2$ denotes the commutator. The proof of this result is highly non trivial and makes use of Character Theory and computation where algebraic computer programs were specially designed.

Once this conjecture was proved, it seems natural to consider Engel words of arbitrary length instead of the commutator τ. That is considering the word $E_m = [...[x,y],y], ..., y]$. Ore's Conjecture can be easily extended to Engel words, is it still true that $G = E_m(G)$ for every finite simple group G and any natural number m?

In Reference [13], a first approach was given for alternating simple groups. The author proved that every element in A_n, with $n \geq 5$, can be written as a product of at most two Engel words of arbitrary length, that is

$$A_n = E_{m_1}(A_n) E_{m_2}(A_n),$$

for any natural numbers $m_1, m_2 \geq 2$ and $n \geq 5$. However, the general case for alternating groups (if $A_n = E_m(A_n)$ for any natural numbers $m \geq 2$ and $n \geq 5$) remains unknown.

In Sections 2 and 3 two new approaches to this problem are presented. First, in Section 2, we study special sequences of Engel words, getting interesting properties about their length. In Section 3 we define a graph, depending on an alternating group and a fixed permutation and study the relation between this graph and the fact of an element $y \in A_n$ being an Engel word of arbitrary length.

In Section 4, we work with an alternating group A_n, $5 \leq n \leq 14$ and a fixed permutation y in A_n. We build a graph related to them in order to empirically study the Engel words $E_m(\cdot, y)$ in A_n.

2. Engel Chains

In this section, we define a particular type of sequence of Engel words and study some of their properties. We also analyze computationally the maximal length of these sequences for small alternating groups.

Let y be a fixed element in the alternating group A_n, $n \geq 5$. For each element $x \in A_n$ we can consider the following sequence of Engel words

$$E^y(x) := \{x, E_1(x,y), E_2(x,y), ...\}.$$

There always exist two integers $1 \leq k_1 < k_2$ such that $E_{k_1}(x,y) = E_{k_2}(x,y)$ in $E^y(x)$. Let us consider the first occurrence of k_1 and k_2 and the set

$$B^y(x) := \{E_{k_1}(x,y), ..., E_{k_2-1}(x,y)\}.$$

Definition 1. *Let x and y be two fixed elements in A_n. The set $E^y(x)$ is called the y-Engel Chain associated to the element x and $B^y(x)$ is called y-Engel Loop associated to the element x.*
The length of the Chain $E^y(x)$ is $l(E^y(x)) = k_2 - 1$ and the length of the loop is $l(B^y(x)) = k_2 - k_1$.

Lemma 1. *Given $n \geq 5$ we have that for every $m \in \mathbb{N}$*

$$E_m(x,y)E_{m+1}(x,y) = E_m(x^y, y).$$

Proof. Given $y \in A_n$, for every $m \in \mathbb{N}$ we have that

$$E_m(x,y)^y = E_m(x^y, y).$$

Then

$$E_m(x^y, y) = E_m(x,y)(E_m(x,y))^{-1} y^{-1} E_m(x,y) y = E_m(x,y) E_{m+1}(x,y).$$

□

That is, the product of two consecutive Engel words in the Chain $E^y(x)$ is an Engel word of the Chain $E^y(x^y)$.

Definition 2. *Given two elements $x, y \in A_n$, with $n \geq 5$, the Engel loop $B^y(x)$ is stable by y-conjugation if $(B^y(x))^y = B^y(x)$.*

We give now a characterization of an y-Engel loop stable by y-conjugation.

Lemma 2. *Given $x, y \in A_n$, with $n \geq 5$, the loop $B^y(x)$ is stable by y-conjugation if an only if for every $E_m(x,y) \in B^y(x)$ we have that*

$$E_m(x,y)E_{m+1}(x,y) \in B^y(x).$$

Proof. It is enough to see that

$$(B^y(x))^y := \{E_m(x,y)^y \mid E_m(x,y) \in B^y(x)\},$$

and that $E_m(x,y)^y = E_m(x^y, y)$.
Applying Lemma 1 we get the result. □

Let G be a group and let us consider an element $g \in G$. From now on, $o(g)$ denotes the order of the element g and $C_G(g)$ denotes the centralizer of g of the group G.

Lemma 3. *Let y be a cycle with maximal length in A_n and $B^y(x)$ a loop stable by y-conjugation. Let us take $\delta \in B^y(x)$, we have that*

1. *For every $z \in C_{A_n}(y)\delta$ we have that $[\delta, y] = [z, y]$.*
2. *For every $z \in \delta C_{A_n}(y)$ we have that $[z, y] \in B^y(x)$.*

Proof. (1) is evident. To prove (2), as $C_{A_n}(y) = \langle y \rangle$ we have that

$$[\delta, y]^y = y^{-1}\delta^{-1}y^{-1}\delta yy = [\delta y, y],$$

and therefore, it is also true for every power of y. □

Using Lemma 3, we know that if we consider an element δ in a loop $B^y(x)$ stable by y-conjugation, every element in the set $C_{A_n}(y)\delta$ produces the same element when it is commuted by y. Note that δ is the only element in the set $C_{A_n}(y)\delta$ which belongs to $B^y(x)$.

Furthermore, every element in the set $\delta C_{A_n}(y)$ belongs to the loop $B^y(x)$ when it is commuted by y and since y is a cycle with maximal length in A_n, we have that

$$\delta C_{A_n}(y) = \{[\delta, y]^{y^j} \mid j = 1, \ldots, o(\delta)\}.$$

Therefore condition (2) in Lemma 3 is necessary and sufficient to guarantee that the loop $B^y(x)$ is stable by y-conjugation.

Let us study the amount of conjugated loops in an alternating group A_p, with p prime.

Lemma 4. *Let p be a prime number and $y \in A_p$ a p-cycle. If the loop $B^y(x)$ is not stable by y-conjugation, it has exactly p conjugated loops in the set $S := \{B^y(x) \mid x \in A_p\}$.*

Proof. Let us consider the following action

$$\phi : \langle y \rangle \times S \longrightarrow S$$
$$(y, B^y(x)) \mapsto (B^y(x))^y$$

We have that $\mid Orb(B^y(x)) \mid$ is exactly $\langle y \rangle / \mid Stab_{\langle y \rangle}(B^y(x)) \mid$.

Then if $B^y(x)$ is not stable by y-conjugation, we have that $Stab_{\langle y \rangle}(B^y(x)) = e$ and then $\mid Orb(B^y(x)) \mid = p$. □

Fix an element y in an alternating group A_n, $n \geq 5$ and consider the loop $B^y(x)$ associated to the element x in A_n, we have that

$$B^y(x) := \{E_{k_1}(x,y), \ldots, E_{k_2-1}(x,y)\},$$

where $E_{k_2}(x,y) = E_{k_1}(x,y)$.

Let us fix z an element in the loop $B^y(x)$, since $z = [\tau, y]$ for some element $\tau \in B^y(x)$, we have that $z = [\sigma\tau, y]$ for every $\sigma \in C_{A_n}(y)$. Therefore, the set of elements that, when commuted by y, produces z as a result is

$$A := \{\sigma\tau \mid \sigma \in C_{A_n}(y)\}.$$

Since z and τ are elements in $B^y(x)$, we have that there is only one element in A which also belongs to $B^y(x)$: τ.

Then, for every element z_1 in the loop $B^y(x)$, we have that there is only one element z_2 in $B^y(x)$ such that, when commuted by y, the result is z_1 and there are $\mid C_{A_n}(y) \mid -1$ elements outside of the loop $B^y(x)$ such that commuted by y gives as a result z_1.

Definition 3. *Given an element y in an alternating group A_n, $n \geq 5$, we define the annihilator of y as the set of elements x in A_n such that there exists $k \in \mathbb{N}$ with $E_k(x,y) = e$. We denote this set by $\mathcal{T}_y$.*

The annihilator of y is the set of elements in A_n whose chain 'finishes' in the identity element. If $\mathcal{C}_y$ defines the set of elements in A_n whose chain goes to a loop different from the identity element, we have that
$$A_n = \mathcal{T}_y \cup \mathcal{C}_y.$$

We also know that $\mathcal{C}_y = \cup_{i=0}^{\alpha(y)} \mathcal{C}_y^i$, where $\mathcal{C}_y^0$ is the set of non-identity elements that belong to the loop $B^y(x)$, $\mathcal{C}_y^1$ is the set of elements in A_n which do not belong to $\mathcal{C}_y^0$ but its commutator with y belongs to $\mathcal{C}_y^0$ and, inductively,
$$\mathcal{C}_y^{i+1} := \{ x \in A_n \setminus \mathcal{C}_y^i \mid [x,y] \in \mathcal{C}_y^i \}.$$

Note that there exists an index $\alpha(y) \in \mathbb{N}$ such that $\mathcal{C}_y^{\alpha(y)+1} = \emptyset$.

We performed a brute-force search (using GAP) to study the length of the y-Engel chains that finishes in the identity element for small alternating groups A_n, $5 \leq n \leq 14$.

The results we obtained were that the maximal length of these chains were 2 or 3 for the considered alternating groups. We summarize all the information in Table 1.

Table 1. Computational results for Engel Chains' length.

Alternating Group	Max. Length
A_5	2
A_6	2
A_7	2
A_8	2
A_9	3
A_{10}	3
A_{11}	2
A_{12}	2
A_{13}	2
A_{14}	2

This output is quite interesting as it seems to indicate that the y-Engel Chains that finish in the identity element are usually short.

In the following lines, we prove that under certain conditions, the maximal length of an Engel Chain that end in the identity element is 2. Note that $N_{A_n}(C_{A_n}(y))$ denotes the normalizer of $C_{A_n}(y)$ in A_n.

Lemma 5. *Let us consider the group A_n, $n \geq 5$ odd and take $y = (1, 2, ..., n)$, a cycle with maximal length in A_n. We have that $C_{A_n}(y) = \langle y \rangle$ and that $\mid N_{A_n}(C_{A_n}(y)) \mid$ is either $n\varphi(n)$ or $n\varphi(n)/2$, where φ is the Euler's totient function.*

Proof. The number of conjugated elements of an n-cycle in S_n is $(1/n)V_n^n = (n-1)!$. Then, $\mid S_n : C_{S_n}(y) \mid = (n-1)!$. We have that
$$\mid C_{S_n}(y) \mid = \frac{\mid S_n \mid}{\mid Cl_{S_n}(y) \mid} = \frac{n!}{(n-1)!} = n.$$

Since $\mid \langle y \rangle \mid = n$, we have that $C_{S_n}(y) = \langle y \rangle$. In the group $\langle y \rangle$, there exist exactly $\phi(n)$ elements with the same decomposition as a product of disjoint cycles as y, so we have that $\mid N_{S_n}(\langle y \rangle) \mid = n\varphi(n)$.

By definition, $N_{A_n}(\langle y \rangle) = \{ x \in A_n \mid y^x \in \langle y \rangle \}$. If for every i with $gcd(i, n) = 1$ we have that the elements y and y^i are conjugated in A_n, the number of groups in A_n conjugated to the group $\langle y \rangle$ would be half of the number of groups in S_n conjugated to $\langle y \rangle$. So $N_{A_n}(\langle y \rangle) = N_{S_n}(\langle y \rangle)$ and then
$$\mid N_{A_n}(\langle y \rangle) \mid = n\varphi(n).$$

If half of the powers of y are conjugated to y in A_n, we have that there exists $\sigma \in S_n \setminus A_n$ such that $\sigma \in N_{S_n}(\langle y \rangle)$. Then,

$$|N_{A_n}(\langle y \rangle)| = \frac{|N_{S_n}(\langle y \rangle)|}{2} = \frac{n\varphi(n)}{2}.$$

□

Lemma 6. *Let p be a prime number greater than 3. Let us consider y, a cycle of maximal length in A_p. The annihilator of y, $\mathcal{T}_y$, in A_p is the group $N_{A_p}(\langle y \rangle)$.*

Proof. Consider $Z = \langle y \rangle = C_{A_p}(y)$ and $N_1 = N_{A_p}(Z)$. We define $N_2 := \{x \in A_n \mid Z^x \subset N_1\}$ and inductively

$$N_r := \{x \in A_n \mid Z^x \subset N_{r-1}\}.$$

Note that $E_3(x, y) = 1$ if and only if $E_2(E_1(x, y), y) = e$, that is, $E_1(x, y) \in N_1$. Then, $x^{-1}y^{-1}xy$ is an element of N_1 and therefore $y \in Z \subset N_1$; $(y^{-1})^x$ is an element of N_1, that is, $Z^x \subset N_1$.

We have proved that $x \in N_2$ if an only if $E_3(x, y) = 1$. We will prove by induction that $E_{r+1}(x, y) = 1$ if an only if $x \in N_r$.

$E_{r+1}(x, y) = 1$ if and only if $E_r(E_1(x, y), y)$ that is (by induction), $E_1(x, y) \in N_{r-1}$ and therefore, $[x, y] \in N_{r-1}$.

Then $(y^{-1})^x \in N_{r-1}$, that is, $Z^x \subset N^{r-1}$ and by definition we have that $x \in N_r$.

We have two chains:

- $Z \subset N_1 \subset N_2 \subset N_3 \subset \ldots$
- $Z \subset N_{A_p}(Z) = N_1 \subset N_{A_p}(N_1) = \tilde{N}_2 \subset N_{A_p}(N_2) = \tilde{N}_3 \subset \ldots$

Since p is a prime number we have that $Z \in Syl_p(A_p)$ and, since N_1 is selfnormalizer, we have that $\tilde{N}_2 = N_1$.

If we take x an element in N_2, we have that $Z^x \subset N_1$ and $Z, Z^x \in Syl_p(N_1)$. Then we have that $Z = Z^x$ and therefore $x \in N_1$. So, $N_1 = N_2$.

Since $\mathcal{T}_y = \cup_{i \geq 1} N_i$, we have that

$$\mathcal{T}_y = N_1 = N_{A_p}(\langle y \rangle).$$

□

Lemma 7. *Let n be a positive integer such that $\gcd(n, \varphi(n)) = 1$ and let y be a n-cycle in A_n. The annihilator of y, $\mathcal{T}_y$, in A_n is the group $N_{A_n}(\langle y \rangle)$.*

Proof. Let p_i be a prime divisor of n. Given $P_i \in Syl_{p_i}(\langle y \rangle)$, we have that $P_i \trianglelefteq N_{A_n}(\langle y \rangle)$ and since $\gcd(n, \varphi(n)) = 1$ we have that $\langle y \rangle$ is the only subgroup of $N_{A_n}(\langle y \rangle)$ with order n.

Using the arguments from Lemma 5, we have that $N_1 = N_2$. □

Corollary 1. *Let n be a positive integer such that $\gcd(n, \varphi(n)) = 1$ and y be a cycle of maximal length in A_n. Then the maximal length of an Engel Chain $E^y(x) \subset \mathcal{T}_y$ which ends in the identity element is 2.*

Proof. If $E_m(x, y) = e$ we have that $E_{m-1}(x, y) \in \langle y \rangle$. Also we have that $E_{m-2}(x, y) \in N_{A_n}(\langle y \rangle)$.

Thanks to Lemma 7, we have that $N_{A_n}(\langle y \rangle)$ is self-normalizing. Then the maximal length of the chain $E^y(x)$ is $m - (m - 2) = 2$. □

3. Engel Graphs

Let y be a fixed element in an alternating group A_n, with $n \geq 5$ and a $m \geq 1$, let us consider the following set of Engel words of length m:

$$E_m(y) := \{E_m(x,y) \mid x \in A_n\}.$$

Since for every $m \geq 1$ we have that $E_{m+1}(y) \subset E_m(y)$, $\{E_m(y)\}_{m \geq 0}$ is a descending chain of subsets in A_n.

Let us fix $m \geq 1$ and consider the set $E_m(y)$ as $\{[x,y] \mid x \in E_{m-1}(y)\}$, where $E_0 = A_n$. Then, if $x, z \in E_{m-1}(y)$ we have that

$$[x,y] = [z,y] \quad \text{if and only if} \quad C_{A_n}(y)x = C_{A_n}(y)z.$$

Let us consider the set $\Omega_m^y := \{C_{A_n}(y)x \mid x \in E_{m-1}(y)\}$. We can define the following map

$$\varphi_m : \begin{array}{ccc} \Omega_m^y & \longrightarrow & E_m(y) \\ C_{A_n}(y)x & \mapsto & [x,y] \end{array} \tag{1}$$

It is easy to see that for every $m \geq 1$ and every element $y \in A_n$, $n \geq 5$, the map φ_m is well defined and bijective.

Then, we can study the sets $E_m(y)$ by working with the set Ω_1^y of all right cosets of $C_{A_n}(y)$ in A_n. Note that as $\{\Omega_m^y\}_{m \geq 1}$ is a descending chain of sets and A_n is a finite group, there exists $m \in \mathbb{N}$ such that $\Omega_m^y = \Omega_{m+1}^y$.

We are going to define a directed graph which will allow us the study of Engel words in A_n. Let us consider the set of nodes $V_n^y := \Omega_1^y$ and let us define the set of arrows $\mathbb{A}$ by the following relation:

- Given $z_1, z_2 \in V_n^y$, there exists an arrow from z_1 to z_2 if an only if $C_{A_n}(y)[z_1, y] = C_{A_n}(y)z_2$.

Definition 4. *Let y be an element in an alternating group A_n, the graph $(V_n^y, \mathbb{A})$ is called Engel graph associated to the element y and the group A_n.*

It is possible to use this graph in the study of Engel words in an alternating group as:

- If we consider a path of length k in the graph, starting in the node $C_{A_n}(y)z_1$ and finishing in the node $C_{A_n}(y)z_{k+1}$, we have that $E_k(z_1, y) = [z_{k+1}, y]$. Once the graph is built, it is possible to easily compute Engel words of high lengths.
- Reciprocally, if we want to compute $E_k(x, y)$, it is enough to consider a path of length k starting in the node $C_{A_n}(y)x$ and commute by y any element of the coset associated to the last node of the path $C_{A_n}(y)z_{k_1}$. We have that

$$E_k(x, y) = [z_{k-1}, y].$$

- We can study the 'dynamic' of the set $\{E_m(\cdot, y)\}_{m \geq 0}$ by studying the 'dynamic' of the graph $(V_n^y, \mathbb{A})$.

Once the graph is constructed, we want to use it to know whether or not an element in the alternating group A_n, $n \geq 5$, can be written as an Engel word of type $E_m(\cdot, y)$ for $m \geq 1$. The following lemma shows the relation between the graph $(V_n^y, \mathbb{A})$ and the fact of an element in the alternating group being an Engel word of arbitrary length.

Lemma 8. *Let φ_1 be the map defined in (1) with $m = 1$. If (W, β) is a directed cycle of $(V_n^y, \mathbb{A})$, every element in the set $\varphi_1(W)$ can be written as an Engel word of arbitrary length.*

Proof. Consider (W, β), a directed cycle in the Engel graph $(V_n^y, \mathbb{A})$.

Fixing an arbitrary element $C_{A_n}(y)x$ in W, we have that

$$\varphi_1(W) := \{E_k(x,y) \mid k \in \mathbb{N}\}.$$

As W is a directed cycle, there exists $k_1 \in \mathbb{N}$ such that $[x,y] = E_{k_1}(x,y)$.

Take an arbitrary $m \in \mathbb{N}$ and a permutation σ in $\varphi_1(W)$. We have that $\sigma = [z,y]$ for $z \in C_{A_n}(y)x$ and there exists $k_2 \in \mathbb{N}$ such that $[z,y] = E_{k_2}(z,y) = E_{2k_2}(z,y) = ...E_{rk_2}(z,y)$, with $r \in \mathbb{N}$.

It is enough to take $k_2 > m$ to get that $\sigma = E_m(\tau, y)$ for some $\tau \in A_n$. □

Lemma 8 implies that given an alternating group A_n, $n \geq 5$ and y an element in A_n, if we compute φ_1 of the directed cycles in the Engel graph we get a subset of A_n in which every element can be written as an Engel word of arbitrary length.

Corollary 2. *If (W, β) is a directed cycle of $(V_n^y, \mathbb{A})$ and φ_1 the map defined in (1) with $m = 1$, every element $\varphi_1(W)^{S_n}$ can be written as an Engel word of arbitrary length in A_n.*

Proof. This result can be directly deduced from Reference [13] and Lemma 8. □

The following results shows some of the properties that Engel Graphs have.

Lemma 9. *If $m \geq n$ and $\phi : A_n \longrightarrow A_m$ is the natural embedding, the image by ϕ of a directed cycle in an Engel graph $(V_n^y, \mathbb{A})$, is a directed cycle in the Engel graph $(V_m^y, \mathbb{B})$.*

Proof. Fix $y \in A_n$ and let W_1 be a directed cycle of the Engel graph $(V_n^y, \mathbb{A})$. Given a node $C_{A_n}(y)x$ of W_1, we can consider the directed cycle W_2 of $(V_m^y, \mathbb{B})$ that contains the node $C_{A_m}(y)x$.

If there exists an arrow between two nodes x, z of W_1, we have that

$$C_{A_n}(y)[x,y] = C_{A_n}(y)z.$$

Then $[x,y]z^{-1} \in C_{A_n}(y) \subset C_{A_m}(y)$ for every $m \geq n$. Then there is an arrow between the nodes $\phi(x)$ and $\phi(y)$ in W_2. As W_1 is a directed cycle, we have that W_2 is also a directed cycle of the same length that W_1. □

Corollary 3. *Every element in A_n that can be written as an Engel word of arbitrary length in A_n, is also an Engel word of arbitrary length in A_m, for every $m \geq n$.*

A sufficient condition for two Engel graphs to be isomorphic is presented in the following result.

Lemma 10. *If $z \in Cl_{S_n}(y)$ we have that the Engel graphs $(V_n^y, \mathbb{A})$ and $(V_n^z, \mathbb{B})$ are isomorphic.*

Proof. Denote $z := y^x$ for some $x \in S_n$. We define the next map

$$\phi : V_n^y \longrightarrow V_n^z$$
$$C_{A_n}(y)\sigma \mapsto C_{A_n}(z)\sigma^x$$

If $C_{A_n}(y)x_1 = C_{A_n}(y)x_2$, we have that $x_1x_2^{-1} \in C_{A_n}(y)$. Then

$$(x_2x_1^{-1})^x y^x (x_1x_2^{-1})^x = y^x,$$

so $C_{A_n}(z)x_1^x = C_{A_n}(z)x_2^x$ and then ϕ is injective.

Surjectivity is obvious, so ϕ is a bijection.

Consider two nodes $C_{A_n}(y)x_1$ and $C_{A_n}(y)x_2$ in $(V_n^y, \mathbb{A})$, such that there is an arrow from $C_{A_n}(y)x_1$ to $C_{A_n}(y)x_2$, that is $C_{A_n}(y)[x_1.y] = C_{A_n}(y)x_2$.

We have that

$$x_2[x_1,y]^{-1}y[x_1,y]x_2^{-1} = y,$$

and then

$$(x_2[x_1,y]^{-1})^x y^x ([x_1,y]x_2^{-1})^x = y^x,$$

so

$$C_{A_n}(z)[x_1^x, z] = C_{A_n}(z)x_2^x.$$

If there is an arrow between two nodes in $(V_n^y, \mathbb{A})$, there is also an arrow between the image of these nodes by ϕ in $(V_n^z, \mathbb{B})$. Then ϕ is a isomorphism of graphs. □

4. Engel Graphs for Small Alternating Groups

In this section, we use an Engel graph to prove that $A_n = E_m(A_n)$ for every $m \geq 1$ and every $n \leq 14$. We show here the explicit method performed for the alternating group A_5. For $6 \leq n \leq 14$, the procedure is analogous and we show the computational results at the end of this section.

Some results from Reference [13] are necessary to prove Theorem 2. We summarize those results in the following lemma.

Lemma 11. *Let $\sigma \in A_n$, $n \geq 5$, be a permutation of one of the following types: a product of two transpositions, a 3-cycle or a product of two 3-cycle. Then σ is an Engel word of arbitrary length in A_n.*

Consider $y := (1,2,3,4,5)$ a 5-cycle in A_5. We have that $C_{A_5}(y) = \langle y \rangle$, the cyclic group of order 5, so $V_5^y = \{\langle y \rangle x \mid x \in A_5\}$ is a set of order $\mid A_5/\langle y \rangle \mid = 12$.

Let us build the Engel graph $(V_5^y, \mathbb{A})$ in Figure 1. As we know, each node is associated to a coset module $C_{A_5}(y)$. We denote each node $C_{A_5}(y)\sigma$ by a permutation of the set $\{y^j\sigma \mid 1 \leq j \leq 4\}$.

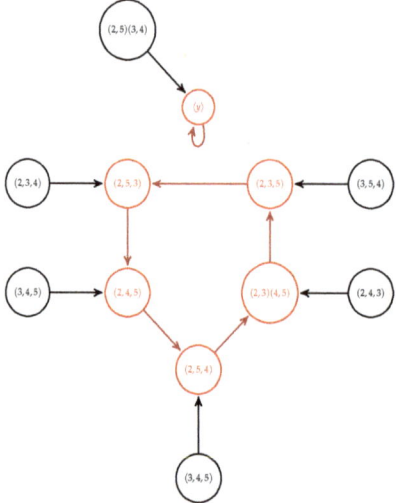

Figure 1. Engel Graph $(V_5^{(1,2,3,4,5)}, \mathbb{A})$.

The graph has two directed cycles. The first one, W_1, is a cycle with five elements and the other one, W_2, is only the identity node, $C_{A_5}(y)$.

By using Lemma 8, we just need to compute the sets $\varphi_1(W_1)$ and $\varphi_1(W_2)$ to get another set of elements in A_5 which can be written as an Engel word of arbitrary length in A_5. We have that:

$$\varphi_1(W_2) := \{e\},$$

$$\varphi_1(W_1) := \{(1,3,2,5,4), (1,3,5,4,2), (1,4,3,5,2), (1,5,2,4,3), (1,5,3,2,4)\}.$$

Thanks to Corollary 2 we have that every 5-cycle in A_5 can be written as an Engel word of arbitrary length in A_5.

This result together with Lemma 11 allows us to prove the following theorem:

Theorem 1. *Every element in A_5 can be written as an Engel word of arbitrary length. That is, for every $n \geq 1$ we have that $A_5 = E_n(A_5)$.*

It is also possible to use the adjacency matrix of the Engel graph to study which nodes belong to a directed cycle. If we consider Λ the adjacency matrix of the Engel graph $(V_n^y, \mathbb{A})$, it is known that the element a_{ij} of the matrix Λ^k gives us the number of directed paths of length k from the node i to the node j in the graph.

Computing the powers of the adjacency matrix and looking for the elements in the diagonal of Λ^k that are different to 0, we can compute which elements of the graph belong to a cycle.

Let us consider the graph $(V_5^y, \mathbb{A})$, with $y := (1,2,3,4,5)$. Its associated adjacency matrix Λ is:

$$\begin{pmatrix}
1 & 0 & 0 & 0 & 0 & 0 & 0 & 0 & 0 & 0 & 0 & 0 \\
1 & 0 & 0 & 0 & 0 & 0 & 0 & 0 & 0 & 0 & 0 & 0 \\
0 & 0 & 0 & 1 & 0 & 0 & 0 & 0 & 0 & 0 & 0 & 0 \\
0 & 0 & 0 & 0 & 1 & 0 & 0 & 0 & 0 & 0 & 0 & 0 \\
0 & 0 & 0 & 0 & 0 & 0 & 1 & 0 & 0 & 0 & 0 & 0 \\
0 & 0 & 0 & 0 & 1 & 0 & 0 & 0 & 0 & 0 & 0 & 0 \\
0 & 0 & 0 & 0 & 0 & 0 & 0 & 1 & 0 & 0 & 0 & 0 \\
0 & 0 & 0 & 0 & 0 & 0 & 0 & 0 & 1 & 0 & 0 & 0 \\
0 & 0 & 1 & 0 & 0 & 0 & 0 & 0 & 0 & 0 & 0 & 0 \\
0 & 0 & 0 & 0 & 0 & 1 & 0 & 0 & 0 & 0 & 0 & 0 \\
0 & 0 & 0 & 0 & 0 & 0 & 0 & 1 & 0 & 0 & 0 & 0 \\
0 & 0 & 0 & 0 & 0 & 0 & 0 & 0 & 1 & 0 & 0 & 0
\end{pmatrix}$$

If we compute Λ^5, the result is:

$$\begin{pmatrix}
1 & 0 & 0 & 0 & 0 & 0 & 0 & 0 & 0 & 0 & 0 & 0 \\
1 & 0 & 0 & 0 & 0 & 0 & 0 & 0 & 0 & 0 & 0 & 0 \\
0 & 0 & 0 & 0 & 0 & 0 & 0 & 0 & 1 & 0 & 0 & 0 \\
0 & 0 & 0 & 1 & 0 & 0 & 0 & 0 & 0 & 0 & 0 & 0 \\
0 & 0 & 0 & 0 & 1 & 0 & 0 & 0 & 0 & 0 & 0 & 0 \\
0 & 0 & 0 & 1 & 0 & 0 & 0 & 0 & 0 & 0 & 0 & 0 \\
0 & 0 & 0 & 0 & 0 & 0 & 1 & 0 & 0 & 0 & 0 & 0 \\
0 & 0 & 0 & 0 & 0 & 0 & 0 & 1 & 0 & 0 & 0 & 0 \\
0 & 0 & 0 & 0 & 0 & 0 & 0 & 0 & 1 & 0 & 0 & 0 \\
0 & 0 & 0 & 0 & 1 & 0 & 0 & 0 & 0 & 0 & 0 & 0 \\
0 & 0 & 0 & 0 & 0 & 0 & 1 & 0 & 0 & 0 & 0 & 0 \\
0 & 0 & 0 & 0 & 0 & 0 & 0 & 1 & 0 & 0 & 0 & 0
\end{pmatrix}$$

It is possible to see that there are 6 nodes in the Engel graph $(V_5^y, \mathbb{A})$ that belong to a directed cycle. However, as the size of the matrix corresponds to the number of nodes in the Engel graph, working with these kinds of matrices becomes impractical when we consider alternating groups of higher order. As an example, for $y = (1,2,3,4,5)$ in A_6 the set V_6^y has $|A_6| / |C_{A_6}(y)| = 72$ elements. For A_7 and y a 7-cycle, we get 360.

To study bigger alternating groups, we used GAP to compute the directed cycles $\{W_k\}$ of the Engel graph associated to the group A_n and the element $y \in A_n$.

Later, we computed the set $\varphi_1(W_k)$ for each directed cycle W_k in the graph $(V_5^y, \mathbb{A})$. Then, we find out which types of permutations belongs to $\cup_k \varphi_1(W_k)$. To finish, we list every type of permutations in A_n that does not belong to $\cup_k \varphi_1(W_k)$.

This final list contains every type of permutation that cannot be written as an Engel word of arbitrary length of type $E_m(x,y)^\sigma = E_m(x^\sigma, y^\sigma)$.

Let us fix a cycle y of maximal length in A_n, $5 \leq n \leq 14$. We use the previous algorithm to search the directed cycles of the Engel graph $(V_n^y, \mathbb{A})$ in order to see if Theorem 1 is also true for bigger alternating groups.

We will compute the set $\Omega := \cup_k \varphi_1(W_k)$, where $\{W_k \mid 1 \leq k \leq r\}$ is the set of directed cycles in the Engel graph and we will see what types of permutations do not appear in Ω.

- Using the algorithm described above in GAP for A_6 and $y = (1,2,3,4,5)$, we get that the types of permutations in A_6 which do not appear in Ω are

$$\{(1,2)(3,4), (1,2,3), (1,2,3)(4,5,6)\}.$$

Applying Lemma 11, we can get Theorem 1 for the group A_6.

- If we take A_n, with $7 \leq n \leq 14$ and we repeat the same process for $y = (1,2,3,...,n)$, if n is odd but $y = (1,2,3,...,n-1)$ if n is even, there is only one type of permutation that does not appear in the set Ω: $\{(1,2)(3,4)\}$.

And again, we can easily get the Theorem 1 for the groups A_n, with $7 \leq n \leq 14$.

We summarise all the results we have got computationally in Table 2.

Table 2. Computational results for Engel graphs

Group	Conj. Cl. Not Found	Run Time
A_5	$\{(1,2)(3,4)^{S_5}, (1,2,3)^{S_5}\}$	7 ms
A_6	$\{(1,2)(3,4)^{S_6}, (1,2,3)^{S_6},$ $(1,2,3)(4,5,6)^{S_6}\}$	18 ms
A_7	$\{(1,2)(3,4)^{S_7}\}$	40 ms
A_8	$\{(1,2)(3,4)^{S_8}\}$	201 ms
A_9	$\{(1,2)(3,4)^{S_9}\}$	4 s 12 ms
A_{10}	$\{(1,2)(3,4)^{S_{10}}\}$	40 s 809 ms
A_{11}	$\{(1,2)(3,4)^{S_{11}}\}$	5 min 37 s 139 m
A_{12}	$\{(1,2)(3,4)^{S_{12}}\}$	63 min 38 s 210 m
A_{13}	$\{(1,2)(3,4)^{S_{13}}\}$	21 h 6 min 54 s
A_{14}	$\{(1,2)(3,4)^{S_{14}}\}$	approx. 12 days

Theorem 2. *Every element in an alternating group A_n, $5 \leq n \leq 14$, can be written as an Engel word of arbitrary length in A_n. That is,*

$$A_n = E_m(A_n),$$

for every $m \geq 1$.

In this work, we have provided two new approaches that can be used in the study of Engel words in alternating groups: Engel chains and Engel graphs. Using them (and GAP), we have also proved that every element in an alternating group A_n, $5 \leq n \leq 14$, can be written as an Engel word of arbitrary length.

It is still unknown whether Theorem 2 holds for $n > 14$. However, computational results seems to indicate some consistency in the "behaviour" of the Engel words in an alternating group and it is possible that a similar theorem holds for any alternating group A_n, $n \geq 5$. The techniques proposed in this paper might be helpful in the further study of the general problem.

Funding: This work has been partially supported by BES-2011-044790 (research fellowship associated to project MTM2010-18370-C04-01) and GRUPIN 14-142.

Conflicts of Interest: The authors declare no conflict of interest.

References

1. Dehn, M. Uber die topologie des dreidimensionalen raumes. *Math. Ann.* **1910**, *69*, 137–168. (In German) [CrossRef]
2. Artin, E. The theory of braids. *Ann. Math.* **1947**, *48*, 101–126. [CrossRef]
3. David B.A.; Epstein, M.S.; Paterson, J.W.; Cannon, D.F.; Holt, S.V.; Levy, W.P. Thurston. In *Word Processing in Groups*, 1st ed.; A K Peters/CRC Press: Boca Raton, FL, USA, 1992.
4. Ko, K.H.; Lee, S.J.; Cheon, J.H.; Han, J.W.; Kang, J.; Park, C. New public-key cryptosystem using braid group. In *Advances in Cryptology—CRYPTO 2000*; Bellare, M., Ed.; Lecture Notes in Computer Science 1880; Springer: Berlin, Germany, 2000; pp. 166–183.
5. Garside, F.A. The braid group and other groups. *Quart. J. Math. Oxf.* **1969**, *20*, 235–254. [CrossRef]
6. Hofheinz, D.; Steinwandt, R. A practical attack on some braid group based cryptographic primitives. In *Public Key Cryptography—PKC2003*; Desmedt, Y.G., Ed.; Lecture Notes in Computer Science 2384; Springer: Berlin, Germany, 2002; pp. 176–189.
7. Anshel, I.; Anshel, M.; Goldfeld, D. An algebraic methodfor public-key cryptography. *Math. Res. Lett.* **1999**, *6*, 287–291. [CrossRef]
8. Ore, O. Some Remarks on Commutators. *Proc. Am. Math. Soc.* **1951**, *2*, 307–314. [CrossRef]
9. Wilson, J.S. *First-Order Group Theory*; Infinite Groups (1994); Gruyter: Berlin, Germany, 1996; pp. 301–314.
10. Martinez, C.; Zelmanov, E.I. Product of powers in finite simple groups. *Isr. J. Math.* **1996**, *96*, 469–479. [CrossRef]
11. Saxl, J.; Wilson, J.S. A note on powers in simple groups. *Math. Proc. Camb. Philos. Soc.* **1997**, *122*, 91–94. [CrossRef]
12. Liebeck, M.W.; O'Brien, E.A.; Shalev, A.; Tiep, P.H. The Ore Conjecture. *J. Eur. Math. Soc.* **2010**, *12*, 939–1008. [CrossRef]
13. Carracedo, J.M. Engel Words in Alternating Groups. *J. Algebra Appl.* **2017**, *16*, 1750021. [CrossRef]

© 2019 by the authors. Licensee MDPI, Basel, Switzerland. This article is an open access article distributed under the terms and conditions of the Creative Commons Attribution (CC BY) license (http://creativecommons.org/licenses/by/4.0/).

Article

MPF Problem over Modified Medial Semigroup Is NP-Complete

Eligijus Sakalauskas [†] and Aleksejus Mihalkovich *

Department of Applied Mathematics, Kaunas University of Technology, LT-44249 Kaunas, Lithuania; eligijus.sakalauskas@ktu.lt
* Correspondence: aleksejus.michalkovic@ktu.lt; Tel.: +370-600-14070
† Current address: Studentu str. 50-324.

Received: 21 September 2018; Accepted: 22 October 2018; Published: 1 November 2018

Abstract: This paper is a continuation of our previous publication of enhanced matrix power function (MPF) as a conjectured one-way function. We are considering a problem introduced in our previous paper and prove that tis problem is NP-Complete. The proof is based on the dual interpretation of well known multivariate quadratic (MQ) problem defined over the binary field as a system of MQ equations, and as a general satisfiability (GSAT) problem. Due to this interpretation the necessary constraints to MPF function for cryptographic protocols construction can be added to initial GSAT problem. Then it is proved that obtained GSAT problem is NP-Complete using Schaefer dichotomy theorem. Referencing to this result, GSAT problem by polynomial-time reduction is reduced to the sub-problem of enhanced MPF, hence the latter is NP-Complete as well.

Keywords: cryptography; non-commutative cryptography; one-way functions; NP-Completeness; key agreement protocol

1. Introduction

It is very natural to look for a new conjectured one-way functions (OWFs) for cryptographic applications in connection with new challenges caused by quantum cryptanalysis. This paper is a continuation of research in this field and is dealing with so called matrix power function (MPF). Some cryptographic primitives were built on the basis that MPF is a conjectured OWF in [1–5]. Furthermore, some results were published considering the security of presented primitives in [6–8]. The security of these primitives is based on the complexity of MPF inversion named as MPF problem.

So far, it is thought that OWF security based on the NP-Complete problem is not vulnerable to the quantum cryptanalysis, while the cryptosystems based on conjectured OWFs such as factoring and discrete logarithm problems are vulnerable due to [9]. Therefore, it is very desirable to try to prove NP-Completeness of MPF problem. In [6] the NP-Completeness of a more general problem named as multivariate quadratic power problem is presented. However, the question of NP-Completeness of MPF problem remained open so far.

In [10] our efforts were directed toward the increasing expectable complexity of MPF problem by choosing more complicated algebraic structures for MPF definition but at the same time preserving the necessary properties for the cryptographic primitives construction. In that paper, we presented a key agreement protocol in Section 2, Construction 1 as well as an example of its realization with artificially small parameters in Section 6.

In this paper we present a proof of NP-Completeness of sub-problem of enhanced MPF problem previously considered in [10]. The notion of sub-problem is defined as follows:

Definition 1. *The decision problem P_1 is a sub-problem of problem P_2 if every assignment to input values, which provides the answer YES to problem P_2, also implies the answer YES to the problem P_1.*

The proof is based on the duality of multivariate quadratic MQ problem interpretation as a system of MQ equations over $\mathcal{Z}_2 = \{0, 1\}$ [11,12] and according to Schaefer dichotomy theorem [13] as a general satisfiability (GSAT) problem.

The main benefit of such approach is the opportunity to include some constraints to MPF necessary to construct cryptographic primitives as an additional GSAT equations.

The proof is based on proving that this GSAT is NP-Complete and on polynomial-time reduction from GSAT to the sub-problem of enhanced MPF problem.

2. Matrix Power Function

MPF was first introduced in [4]. To be self-contained, we present here MPF in the following way:

Definition 2. *Symbolically MPF corresponds to matrix $W_{m \times m} = \{w_{ij}\}$ powered by matrix $X_{m \times m} = \{x_{ij}\}$ on the left and by matrix $Y_{m \times m} = \{y_{ij}\}$ on the right with MPF value equal to matrix $E_{m \times m} = \{e_{ij}\}$ and is expressed in the following way*

$$^X W^Y = E, e_{ij} = \prod_{k=1}^{m} \prod_{l=1}^{m} w_{kl}^{x_{ik} \cdot y_{lj}}. \tag{1}$$

The matrix W that is powered is named the base matrix and the matrices X and Y that are powering the base matrix are named power matrices. In general, we define the base matrix over the multiplicative (semi)group $\mathcal{S}$ and power matrices over some numerical (semi)ring $\mathcal{R}$. We call semigroup $\mathcal{S}$ a platform (semi)group, which according to the MPF definition, is multiplicative, and $\mathcal{R}$–an exponent (semi)ring. The appropriate matrix semigroups $\mathcal{M}_\mathcal{S}$ and matrix semiring $\mathcal{M}_\mathcal{R}$ contain base matrices and power matrices respectively.

The exact MPF definition depends on the type of sets over which matrices are defined.

In [3] authors proved, that if platform semigroup and power semiring are commutative, then the following associative properties of MPF takes place:

Definition 3. *MPF is one-side associative, (left-side and right-side associative, respectively) if the following identities hold:*

$$\begin{aligned} ^Y\left(^X W\right) &= {}^{(YX)}W = {}^{YX}W; \\ \left(W^X\right)^Y &= W^{(XY)} = W^{XY}. \end{aligned} \tag{2}$$

Definition 4. *MPF is two-side associative if the following identities hold:*

$$\left(^X W\right)^Y = {}^X\left(W^Y\right) = {}^X W^Y. \tag{3}$$

In [3] authors proved, that if platform semigroup $\mathcal{S}$ and power semiring $\mathcal{R}$ are commutative, then $\text{MPF}_\mathcal{S}^\mathcal{R}$ is one and two-side associative.

It follows from Equation (1), that in general, MPF is a function

$$\text{MPF}: \mathcal{M}_\mathcal{R} \times \mathcal{M}_\mathcal{S} \times \mathcal{M}_\mathcal{R} \mapsto \mathcal{M}_\mathcal{S}.$$

Definition 5. *The direct MPF value computation is to find matrix E, when matrices X, W, Y are given.*

Definition 6. *The inverse MPF value computation is to find matrices X and Y, when matrices W and E are given.*

Definition 7. *MPF problem is its inverse value computation.*

Definition 8. *MPF presented in 1 is a candidate one-way function (OWF) if the following necessary (but not sufficient) conditions are satisfied:*

1. The direct MPF value computation is easy;
2. The MPF problem is polynomially equivalent to a certain hard problem with not known polynomial time algorithm.

Assume, that the base matrix W in Expression 1 is defined over a platform semigroup denoted by $\mathcal{S}$ and the power matrices X and Y are defined over a power semiring denoted by $\mathcal{R}$. We denote the MPF problem defined by these structures by $\text{MPF}_\mathcal{S}^\mathcal{R}$. Assume, that power matrices X and Y have to satisfy some constrains denoted by $\mathcal{C}$. In this case we denote the MPF problem by $\text{MPF}_\mathcal{S}^{\mathcal{R},\mathcal{C}}$.

To build cryptographic primitives, e.g., key agreement protocol, based on $\text{MPF}_\mathcal{S}^\mathcal{R}$ the following additional property must be satisfied: square matrices of m-th order X and Y defined over the power semiring $\mathcal{R}$ must be elements of two subsets $\mathcal{M}_{\mathcal{R},1}$ and $\mathcal{M}_{\mathcal{R},2}$ of commuting matrices in $\mathcal{M}_\mathcal{R}$ respectively, i.e., for any $U \in \mathcal{M}_{\mathcal{R},1}$ and $V \in \mathcal{M}_{\mathcal{R},1}$ the following identities take place

$$\mathcal{C}: \begin{matrix} XU = UX; \\ YV = VY. \end{matrix} \qquad (4)$$

This defines a constrained MPF that we previously denoted by $\text{MPF}_\mathcal{S}^{\mathcal{R},\mathcal{C}}$. Further we will use the single subset of commuting matrices in $\mathcal{M}_\mathcal{R}$, namely the subset of circulant matrices i.e., matrices of the following general form [14]:

$$X = \begin{pmatrix} x_1 & x_m & \cdots & \cdots & x_2 \\ x_2 & x_1 & x_m & \cdots & \cdots \\ x_3 & x_2 & x_1 & \cdots & \cdots \\ \cdots & \cdots & \cdots & \cdots & x_m \\ x_m & \cdots & x_3 & x_2 & x_1 \end{pmatrix}. \qquad (5)$$

Any circulant matrix X can be represented by its column vector $\vec{x}$, which transposed form is expressed by the following row vector $\vec{x}^T = (x_1, x_2, ..., x_m)$. If $\text{MPF}_\mathcal{S}^{\mathcal{R},\mathcal{C}}$ satisfies the conditions of Definition 8, then the following secret-key agreement protocol can be executed as proposed in [10]:

Both parties agree on a public information: the modified medial semigroup $\mathcal{S}$ and a public base matrix W with its entries randomly chosen from $\mathcal{S}$. Alice and Bob can agree on a common key as follows:

1. Alice chooses two secret circulant matrices X and Y at random of size m. Using these matrices she computes the MPF value $A = {}^X W^Y$ and sends it to Bob;
2. Bob chooses two secret circulant matrices U and V at random of size m. Using these matrices he computes the MPF value $B = {}^U W^V$ and sends it to Alice;
3. Alice and Bob compute the same secret key in the following way:

$$K_A = {}^X B^Y = {}^X \left({}^U W^V\right)^Y = {}^U \left({}^X W^Y\right)^V = K_B = K. \qquad (6)$$

The Identity (6) is true due to the fact, that circulant matrices are commuting and associativity Conditions (2) and (3).

Remark 1. *In general two-sided association Condition (3) will be not necessary, if we agree upon on the order of operations, e.g., from the left to the right.*

In our previous research the base matrix W was defined over the multiplicative platform group $\mathcal{Z}_p^* = \{1, 2, ..., p-1\}$ and power matrices X and Y over the numerical power ring $\mathcal{Z}_{p-1} =$

$\{0, 1, 2, ..., p-2\}$. This kind of MPF is denoted by $\text{MPF}_{\mathcal{Z}_p^*}^{\mathcal{Z}_{p-1}}$ and constrained version by $\text{MPF}_{\mathcal{Z}_p^*}^{\mathcal{Z}_{p-1}, \mathcal{C}}$. It represents the MPF defined over commutative algebraic structures considered in [1,2,5,7,15].

However, recently a linear algebra attack to the protocol presented in [3] based on $\text{MPF}_{\mathcal{Z}_p^*}^{\mathcal{Z}_{p-1}, \mathcal{C}}$ was found by [16]. This attack to $\text{MPF}_{\mathcal{Z}_p^*}^{\mathcal{Z}_{p-1}, \mathcal{C}}$ problem runs in polynomial time and hence can be used to break the algorithms presented in [1,3]. The authors of [16] also suggested some improvements of our protocols to resist the proposed attack. In [7] we fixed this flaw for the asymmetric encryption protocol, presented in [1].

The intriguing idea was to extend MPF construction to non-commutative algebraic structures, namely $\mathcal{S}$ and $\mathcal{R}$, hence expecting higher complexity of MPF problem and achieving a higher potential security for the construction of cryptographic primitives. The main problem of this approach was the loss of associativity of MPF, which made its application in cryptography impossible.

This approach was successful and is presented in [10], when platform semigroup $\mathcal{S}$ is a modified medial semigroup and power semiring is a special kind of so called near semiring NSR. In this study as a power semiring we use a semiring of non-negative integers denoted by $\mathcal{N}^0 = \{0, 1, 2, 3, ...\}$. So we deal with the MPF denoted by $\text{MPF}_\mathcal{S}^{\mathcal{N}^0}$. If power matrices satisfies commutation Constraints in (4), then we denote corresponding MPF by $\text{MPF}_\mathcal{S}^{\mathcal{N}^0, \mathcal{C}}$.

In this paper we consider a class of $\text{MPF}_\mathcal{S}^{\mathcal{N}^0, \mathcal{C}}$ problems when power matrices are circulant matrices over the $\mathcal{N}^0$ and hence they are commuting and satisfying Conditions (4). Interestingly enough, matrices X and Y are almost never invertible due to the fact, that both fractions and negative numbers are not contained in $\mathcal{N}^0$. This is essential to our proof of NP-Completeness of the $\text{MPF}_\mathcal{S}^{\mathcal{N}^0, \mathcal{C}}$ problem.

In earlier work, the proof that random generated multivariate quadratic power problem over $\mathcal{Z}_n$ is NP-Complete is presented. This proof is insufficient to prove the NP-Completeness of $\text{MPF}_\mathcal{S}^{\mathcal{N}^0, \mathcal{C}}$ problem due to fact that we are considering a partial case of this problem. Our multivariate quadratic power system of equations is predetermined by the matrix power equations. Hence this special case is not random generated. Therefore, the aim of this paper is to fill this gap.

In general, it is hard to prove that a problem with arbitrary constraints is NP-Complete (NP-Hard). We present here an approach to prove it based on Schaefer dichotomy theorem [13]. This theorem is formulated for the GSAT problem, represented by arbitrary finite set of Boolean relations (formulas) with respect to the finite set of Boolean variables. The theorem defines six criteria when either GSAT is in P or in NP-Complete complexity class.

In this paper, we construct a certain sub-problem of GSAT problem which is a one-to-one mapping of certain sub-problem of $\text{MPF}_\mathcal{S}^{\mathcal{N}^0, \mathcal{C}}$ problem. We show, that this GSAT problem satisfies the Schaefer criteria to be NP-Complete. Hence, using polynomial-time reduction, we will prove that decision version of $\text{MPF}_\mathcal{S}^{\mathcal{N}^0, \mathcal{C}}$ problem is also NP-Complete.

We revise the definition and basic properties of modified medial semigroup in the next section and present the main result in Section 4.

3. Modified Medial Semigroup as Platform Semigroup of MPF

Let us consider medial semigroup $\mathcal{S}_\mathcal{M}$, which was previously introduced by [17]. Assume, that the presentation of this semigroup consists of two generators a and b and a relation $R_\mathcal{M}$ written in the following way:

$$\mathcal{S}_\mathcal{M} = \langle a, b | R_\mathcal{M} \rangle; \tag{7}$$

$$R_\mathcal{M} : \omega_1 ab \omega_2 = \omega_1 ba \omega_2. \tag{8}$$

where ω_1 and ω_2 are arbitrary non-empty words in $\mathcal{S}_\mathcal{M}$, written in terms of generators a and b.

Let us now present an important identity, which is useful to us for application of medial semigroup $\mathcal{S}_\mathcal{M}$ to MPF:

$$(\omega_1 \omega_2)^e = \omega_1^e \omega_2^e. \tag{9}$$

This identity is based on the Relation (8) and is valid for all words $\omega_1, \omega_2 \in S_M$ and any exponent $e \in \mathcal{N}^0$.

To prevent the growth of powers of generators when exponentiation takes place we introduce a modified medial semigroup S with two extra relations R_1 and R_2 in the following general form:

$$R_1 : ba^{p+2}b^{p+1} = ba^2b;$$
$$R_2 : ab^{p+2}b^{p+1} = ab^2a.$$
(10)

Thus, modified medial semigroup S has the following presentation:

$$S = \langle a, b | R_M, R_1, R_2 \rangle,$$
(11)

with relations R_M, R_1 and R_2 defined above.

Note, that we define S as a multiplicative, non-commuting, non-cancellative and infinite semigroup which is a non-symmetric algebraic structure.

Remark 2. *The modified medial semigroup is well defined if relations R_1 and R_2 are symmetric, i.e., they link both generators in such a way, that the order of generators is symmetric and exponents of each generator add up to the same number. In our case the sum of exponents of generators a and b on the left side of R_1 and R_2 in Realtions (10) equals $p+2$ and on the right side it equals 2.*

Remark 3. *In our previous paper we considered a special case of $p = 3$.*

Semigroups S_M and S are made monoids by introducing an empty word as a multiplicatively neutral element, denoted by 1. Then conveniently, the following identities hold for all $\omega \in S_M$:

$$\omega 1 = 1\omega = \omega, w^0 = 1, 0 \in \mathcal{N}^0.$$
(12)

The normal form for the words in S_M was also defined in the following way:

Definition 9. *The normal form $\omega_{M,nf}$ of any word ω_0 in semigroup S_M is expressed as follows:*

$$\omega_{M,nf} = \max_{\alpha_a, \beta_b} b^{\beta_b} a^{r_a} b^{s_b} a^{\alpha_a} = b^{\beta} a^{i_a} b^{j_b} a^{\alpha},$$
(13)

where $\alpha, \beta \in \{0,1\}$ and $\alpha_a, \beta_b, r_a, s_b, i_a, j_b \in \mathcal{N}$.

To obtain the normal form for the word ω we consider its first and last literals. Using Relation (9) we can determine the values of α and β. For example the normal form for the word $b^7 a^8 b^2 a^6$ is $ba^{13} b^8 a$. The word $b^6 a^7 b^3 a^7$ has the same normal form and hence we consider all these words equivalent. The normal form for the word $a^7 b^8 a^2 b^6$ is $b^0 a^9 b^{14} a^0$. Hence in the last case we have $\alpha = 0$ and $\beta = 0$. Evidently for the normal form of the word $a^5 b^7 a^3$ we have $\alpha = 1$ and $\beta = 0$ whereas in case of the word $b^5 a^7 b^3$ we have $\alpha = 0$ and $\beta = 1$. In fact, the normal forms for the presented words are $b^0 a^7 b^7 a$ and $ba^7 b^7 a^0$ respectively. We generally omit zeroth powers when writing normal forms.

On the base of $\omega_{M,nf}$ the normal form in S is defined as follows:

Definition 10. *The normal form ω_{nf} of any word ω_0 in semigroup S is expressed by the following expression:*

$$\omega_{nf} = \min_{i_a, j_b} \max_{\beta, \alpha} b^{\beta} a^{i_a} b^{j_b} a^{\alpha}.$$
(14)

Let T be an additive non-commuting semigroup consisting of the tuples (β, i, j, α), where $\alpha, \beta \in \{0,1\} \subset \mathcal{N}^0$ and $i, j \in \mathcal{N}^0$, with the following addition operation:

$$(\beta_1, i_1, j_1, \alpha_1) + (\beta_2, i_2, j_2, \alpha_2) =$$

$$= (\beta_1, i_1 + \alpha_1 + i_2, j_1 + \beta_2 + j_2, \alpha_2),$$

then there is an isomorphism $\varphi : \mathcal{S}_{\mathcal{M},nf} \mapsto T$, which can be expressed by the following relation for any word ω_{nf}

$$\varphi(\omega_{nf}) = \varphi(b^\beta a^i b^j a^\alpha) = (\beta, i, j, \alpha). \tag{15}$$

Hence, using our notation, we defined $\text{MPF}_\mathcal{S}^{\mathcal{N}^0}$, where $\mathcal{S}$ is modified medial semigroup. It is important to note, that $\text{MPF}_\mathcal{S}^{\mathcal{N}^0}$ satisfies associativity conditions in Definitions (2) and (3) due to the properties of medial semigroup.

Adding the commutation Constraints (4) to the power matrices X and Y defined over $\mathcal{N}^0$, constrained $\text{MPF}_\mathcal{S}^{\mathcal{N}^0}$ problem we denoted by $\text{MPF}_\mathcal{S}^{\mathcal{N}^0, \mathcal{C}}$.

In the next section we prove, that $\text{MPF}_\mathcal{S}^{\mathcal{N}^0, \mathcal{C}}$ problem is NP-Complete.

4. Proof of NP-Completeness

Let us consider the following binary matrix equation:

$$XQY = A, \tag{16}$$

where all matrices Q, A, X and Y are defined over the field $\mathcal{Z}_2 = \{0, 1\}$ with multiplication operation denoted by $\wedge$ (logical AND) and addition operation by $\oplus$ (logical XOR). This equation corresponds to binary matrix multivariate quadratic (BMMQ) equation and associated problem to BMMQ problem.

Definition 11. *The binary matrix MQ (BMMQ) problem is to find matrices X and Y in Equation (16), when matrices Q and A are given.*

Remark 4. *Throughout this paper we assume, that matrix Q is well-balanced, i.e., the quantity of 1's is close to $m^2/2$. Furthermore all the 1's are distributed uniformly in the rows and columns of matrix Q.*

If at least one of square matrices X or Y is invertible, then BMMQ Problem (16) is solvable in polynomial time due to one the following transformations:

$$\begin{aligned} XQ \oplus AY^{-1} &= 0; \\ QY \oplus X^{-1}A &= 0, \end{aligned} \tag{17}$$

since XOR operation is inverse to itself.

It is clear, that both transformations represent the system of m^2 homogeneous linear equations with $2m^2$ unknown variables.

However, if both binary matrices X and Y are singular, then Transformations (17) are not possible and hence the initial Problem (16) bears a resemblance to the well known multivariate quadratic (MQ) problem. It is known, that random generated MQ problem is NP-Complete over any field [11,12].

Hence, we define the following problem:

Definition 12. *The singular binary matrix MQ problem (SBMMQ) is to solve BMMQ problem, when matrices X and Y in Equation (16) are singular.*

It is important to note, that we are interested in this particular problem, since in case of $\text{MPF}_\mathcal{S}^{\mathcal{N}^0, \mathcal{C}}$ power matrices are defined over the semiring $\mathcal{N}^0$ and hence any randomly chosen power matrix is not invertible with overwhelming probability. Here and onwards we say that a random event happens with overwhelming probability if its probability of failure is negligible.

We begin from the complexity consideration of CSBMMQ problem.

Our proof is based on Schaefer dichotomy theorem [13]. Let us define a set of Boolean relations $\{r_1, r_2, \ldots, r_M\}$ with variables defined by two vectors $\vec{x}^T = (x_1, x_2, \ldots, x_m)$ and $\vec{y}^T = (y_1, y_2, \ldots, y_m)$. Then the following generalized satisfiability problem GSAT can be formulated:

$$\begin{cases} r_1(\vec{x}, \vec{y}) = 1; \\ r_2(\vec{x}, \vec{y}) = 1; \\ \ldots \\ r_M(\vec{x}, \vec{y}) = 1, \end{cases} \quad (18)$$

where 1 is a true value assignment to the relations.

Definition 13. *The decision GSAT problem is to answer YES/NO to the question: are there any assignment to the variables $\vec{x}$ and $\vec{y}$ that all Boolean relations in Problem (18) are true?*

Theorem 1. *(Schaefer dichotomy theorem [13]). If at least one of the following criteria is satisfied, then the satisfiability problem GSAT is in P, otherwise it is NP-Complete :*

(a) Every relation in S is satisfied when all the variables are 0 (0-valid clause);
(b) Every relation in S is satisfied when all the variables are 1 (1-valid clause);
(c) Every relation in S is definable by a CNF formula in which each conjunct has at most one negated variable (dual Horn clause);
(d) Every relation in S is definable by a CNF formula in which each conjunct has at most one unnegated variable (Horn clause);
(e) Every relation in S is definable by a CNF formula having at most two literals in each conjunct (bijunctive clause);
(f) Every relation in S is the set of solutions of a system of linear equation over the two element field $\{0,1\}$ (affine clause).

As it was mentioned above, to satisfy the commutation Conditions (4), matrices X and Y are chosen to be circulant. Then matrix Equation (16) can be transformed to the following system of equations:

$$\begin{cases} \vec{x}^T Q_{11} \vec{y} = a_{11}; \\ \vec{x}^T Q_{12} \vec{y} = a_{12}; \\ \ldots \\ \vec{x}^T Q_{mm} \vec{y} = a_{mm}, \end{cases} \quad (19)$$

where vectors $\vec{x}^T$ and $\vec{y}^T$ are row vectors of the first row and first column of matrix Q respectively, and matrices $Q_{11}, Q_{12}, \ldots, Q_{mm}$ are obtained by cyclic permutations of matrix Q. For example, $Q_{11} = Q$ and $Q_{12} = \begin{pmatrix} \vec{q}_2 & \vec{q}_3 & \cdots & \vec{q}_m & \vec{q}_1 \end{pmatrix}$, where the vector $\vec{q}_j$ denotes the j-th column of matrix Q. All matrices Q_{ij} are obtained from the initial matrix by performing shifts of rows and/or columns.

The latter system consist of m^2 quadratic equations with $2m$ variables being a components of vectors $\vec{x}$ and $\vec{y}$. System (19) is a special type of random generated MQ problem over $\mathcal{Z}_2$ defined by special type of matrices $Q_{11}, Q_{12}, \ldots, Q_{mm}$, generated by deterministic permutations of random generated matrix Q in Equation (16). Every equation in System (19) represents a Boolean relation written in terms of logical operations AND and XOR.

To choose a suitable GSAT problem to prove NP-Completeness of the initial $MPF_S^{\mathcal{N}^0, \mathcal{C}}$ problem the set of logical Relations (18) must be supplemented by logical relations defining the singularity constraints of matrices X and Y. Since System (19) is defined over $\mathcal{Z}_2 = \{0,1\}$, these constraints can be expressed by the following Boolean relations:

$$\begin{aligned} \det X &= 0; \\ \det Y &= 0, \end{aligned} \quad (20)$$

where 0 is a false value assignment to the relations. The actual expressions of (20) are determined by the format of matrices X and Y. Hence, here and onwards we consider square matrices of m-th order X and Y with even values of determinants.

Definition 14. *The constrained singular binary matrix MQ problem (CSBMMQ) is to solve SBMMQ problem, when matrices X and Y in Equation (16) are singular and hence satisfy Conditions (4) and (16) while also satisfying Condition (20).*

Theorem 2. *Decision CSBMMQ problem is NP-Complete.*

Proof. To prove the theorem, we use the Schaefer dichotomy theorem. System of binary Equation (19) and Relations (20) represent the system of generalized satisfiability relations in Problem (18) and corresponds to GSAT problem with $M = m^2 + 2$. Then to prove NP-Completeness of CSBMMQ we need to verify inconsistency of Schaefer criteria (a)–(f).

The first two criteria (a) and (b) are not satisfied due to the fact, that we are choosing matrix Q at random and hence the satisfiability of these criteria has a negligible probability.

To verify Schaefer criteria (c)–(e) we denote three pairs of vectors satisfying Equations (19) and (20) by $(\vec{x}_1, \vec{y}_1)$, $(\vec{x}_2, \vec{y}_2)$ and $(\vec{x}_3, \vec{y}_3)$. Note, that we generate circulant matrices from selected vectors to check the validity of Equation (20). Schaefer criteria (c)–(e) can be reformulated as follows [18]:

(c′)   For all pairs $(\vec{x}_1, \vec{y}_1)$ and $(\vec{x}_2, \vec{y}_2)$, satisfying System (19) and Equation (20), the pair $(\vec{x}_1 \vee \vec{x}_2, \vec{y}_1 \vee \vec{y}_2)$ is a solution of System (19) and Equation (20);

(d′)   For all pairs $(\vec{x}_1, \vec{y}_1)$ and $(\vec{x}_2, \vec{y}_2)$, satisfying System (19) and Equation (20), the pair $(\vec{x}_1 \wedge \vec{x}_2, \vec{y}_1 \wedge \vec{y}_2)$ is a solution of System (19) and Equation (20);

(e′)   For all pairs $(\vec{x}_1, \vec{y}_1)$, $(\vec{x}_2, \vec{y}_2)$ and $(\vec{x}_3, \vec{y}_3)$, satisfying System (19) and Equation (20), the pair $((\vec{x}_1 \vee \vec{x}_2) \wedge (\vec{x}_1 \vee \vec{x}_3) \wedge (\vec{x}_2 \vee \vec{x}_3), (\vec{y}_1 \vee \vec{y}_2) \wedge (\vec{y}_1 \vee \vec{y}_3) \wedge (\vec{y}_2 \vee \vec{y}_3))$ is a solution of System (19) and Equation (20).

Remark 5. *All logical operations in criteria (c′)-(e′) are performed component-wise.*

Then applying criterion (c′) to the single equation in System (19) in vector form and assigning arbitrary values to the vectors $(\vec{x}_1, \vec{y}_1)$, $(\vec{x}_2, \vec{y}_2)$ we obtain the corresponding values b_{ij} satisfying the following equation in every case

$$(\vec{x}_1 \vee \vec{x}_2)^T Q_{ij} (\vec{y}_1 \vee \vec{y}_2) = b_{ij}.$$

Evidently, in most cases $b_{ij} \neq a_{ij}$. Note, however, that for this criterion to be valid the identity $b_{ij} = a_{ij}$ has to hold for all $i, j = 1, 2, \ldots, m$. Hence, dual Horn clause in System (19) is not satisfied and criterion (c′) is inconsistent.

Analogously, verifying Horn clause we obtain

$$(\vec{x}_1 \wedge \vec{x}_2)^T Q_{ij} (\vec{y}_1 \wedge \vec{y}_2) = c_{ij},$$

where $c_{ij} \neq a_{ij}$. Hence, Horn clause in System (19) is not satisfied for all $i, j = 1, 2, \ldots, m$ and criterion (d′) is inconsistent.

Inconsistency of criterion (e′) follows directly from the latter three expressions. Note, that the key point which allows us to claim the desired result is Remark 5 since no distributive law can be applied to the latter two expressions.

Criterion (f) is not satisfied since, in general, relations in System (19) are non-linear.

So, CSBMMQ problem is NP-Complete. □

Remark 6. *Two additional Relations (20) are needed to ensure that matrices X and Y are singular and hence to ensure the inconsistency of Schaefer criteria.*

Now we turn to constrained singular matrix multivariate quadratic (CSMMQ) problem defined over the semiring of integers $\mathcal{N}_0$ which we denote by $\text{CSMMQ}_{\mathcal{N}_0}$. This means that Equation (16) and corresponding Conditions (19) and (20) are defined over $\mathcal{N}_0$.

Theorem 3. *CSBMMQ problem is a sub-problem of $\text{CSMMQ}_{\mathcal{N}_0}$.*

Proof. Let us consider all matrices in Equation (16) defined over $\mathcal{N}_0$. Then they can be rewritten in the following way:
$$X = 2U + X';$$
$$Y = 2V + Y';$$
$$Q = 2P + Q';$$
$$A = 2T + A'.$$

By substituting these expressions in Equation (16) we obtain the following result:
$$(2U + X')(2P + Q')(2V + Y') = 2T + A'$$

and hence
$$X'Q'Y' \equiv A' \bmod 2.$$

Let us consider the following decision problem: does there exist assignments to matrices X and Y defined over the semiring $\mathcal{N}_0$ satisfying Equation (16), which adding commutation constraints corresponds to Relations (19), (20) and is a $\text{CSMMQ}_{\mathcal{N}_0}$ problem? Assume, that we have an answer YES to decision $\text{CSMMQ}_{\mathcal{N}_0}$ problem. Due to penultimate equation, it implies the answer YES to CSBMMQ problem.

In computational $\text{CSMMQ}_{\mathcal{N}_0}$ version its transformation to CSBMMQ requires the reduction of the solution modulo 2. This is done in polynomial time.

We proved, that CSBMMQ problem is a sub-problem of $\text{CSMMQ}_{\mathcal{N}_0}$ problem, when semiring $\mathcal{N}_0$ is homomorphically mapped to the field $\mathcal{Z}_2$. □

Since Theorem 3 is valid, every solution of $\text{CSMMQ}_{\mathcal{N}_0}$ problem has to satisfy CSBMMQ problem as well. Clearly, this problem is non-trivial and was proven to be NP-Complete.

Let us consider the following system of equations

$$\begin{cases} X\Lambda Y = B; \\ X\Sigma Y \equiv C \bmod (2p). \end{cases} \quad (21)$$

where p is an odd prime, matrices X, Y, Σ and C are defined over the semiring of positive integers $\mathcal{N}^0$, and matrices Λ and B over the ring $\mathcal{Z}$. Furthermore, the parity of matrices Λ and Σ is the same, i.e., $\Lambda - \Sigma = 2T$, where $T \in \mathcal{M}_{\mathcal{Z}}$.

Theorem 4. *The decision CSMMQ problem, defined by System (21), is NP-Complete.*

Proof. It is easy to assume also with overwhelming probability, that matrices X and Y defined over the $\mathcal{N}^0$ are not invertible. We define the following sub-problem of Problem (21) by reducing its first equation modulo $2p$:

$$\begin{cases} X\Lambda Y \equiv B \bmod (2p); \\ X\Sigma Y \equiv C \bmod (2p). \end{cases} \quad (22)$$

Clearly, if the answer to the initial Problem (21) is YES, then the same answer applies also to Problem (22), since to obtain the solution of the Problem (21) extra matrices T and S in the relations

$$X = (2p)T + \tilde{X}_{2p};$$

$$Y = (2p)S + \tilde{Y}_{2p}$$

have to be found. Here matrices $\tilde{X}_{2p}$ and $\tilde{Y}_{2p}$ satisfy the Problem (22).

We can rewrite the System (22) in the following way by using Chinese Remainder Theorem:

$$\begin{cases} X\Lambda Y \equiv B \bmod p; \\ X\Sigma Y \equiv C \bmod p. \end{cases} \quad (23)$$

$$\begin{cases} X\Lambda Y \equiv B \bmod 2; \\ X\Sigma Y \equiv C \bmod 2; \end{cases} \quad (24)$$

It is important to note, that, due to Chinese Remainder Theorem, Systems (23) and (24) must be considered separately. These systems of equations provide two different and mutually independent components of solution of Problem (22). Matrices $\tilde{X}_{2p}$ and $\tilde{X}_{2p}$ satisfying System (22) are calculated as follows:

$$\tilde{X}_{2p} = p\tilde{X}_2 + (p+1)\tilde{X}_p;$$

$$\tilde{Y}_{2p} = p\tilde{Y}_2 + (p+1)\tilde{Y}_p,$$

where matrices $\tilde{X}_p$ and $\tilde{Y}_p$ satisfy System (23) and $\tilde{X}_2$ and $\tilde{Y}_2$ satisfy System (24).

We can assume, that solution of (23) can be found in polynomial time if at least one of matrices X or Y are invertible modulo p. However, nevertheless we cannot recover the solution of (22) from the one component $(\tilde{X}_p, \tilde{Y}_p)$, i.e., the component $(\tilde{X}_2, \tilde{Y}_2)$ is required. It is directly implied by the Chinese Remainder Theorem isomorphism.

Furthermore, since matrices Λ and Σ have the same parity the following congruence is valid:

$$\Lambda \equiv \Sigma \bmod 2.$$

Hence we have $B \equiv C \bmod 2$, since otherwise the answer to Problem (22) is NO. However in this case we can remove either one of equations of System (24) and hence we obtain a CSBMMQ problem. This problem was proven to be NP-Complete in Theorem 2.

We have shown, that the proof of complexity of Problem (21) relies on the complexity of CSBMMQ problem. Since CSBMMQ is NP-Complete and is a sub-problem of CSMMQ Problem (21), then the latter is also NP-Complete. □

Remark 7. *Theorem 3 is the key factor, which allows us to claim the correctness of Theorem 4. However, based on our logic presented here, we cannot claim, that the singular MMQ problem is NP-Complete over $\mathcal{Z}_p$, where p is prime, due to the fact that CSBMMQ problem is not a sub-problem of the latter problem.*

To demonstrate the relation of CSMMQ Problem (21) to modified medial semigroup $\mathcal{S}$ let us define the following mappings:

$$\lambda\left(b^\beta a^i b^j a^\alpha\right) = (i + \alpha) - (j + \beta); \quad (25)$$

$$\sigma\left(b^\beta a^i b^j a^\alpha\right) = (i + \alpha) + (j + \beta). \quad (26)$$

Remark 8. *Obviously Mappings (25) and (26) define functions of powers i and j if we preset the values of α and β.*

Remark 9. *In general we have $\lambda(w) \in \mathcal{Z}$ and $\sigma(w) \in \mathcal{N}^0$. Furthermore, if $\sigma(w) = 0$, then w is an empty word, i.e., $w = 1$.*

It is clear that if we preset two exponents $\alpha, \beta \in \{0,1\}$, then the pair $(\lambda(w), \sigma(w))$ defines a unique element w if these elements have the same parity and satisfy inequality $|\lambda(w)| < \sigma(w)$. Clearly, this reduction is polynomial since for a fixed pair $\varphi_{(\alpha_0, \beta_0)}(\lambda, \sigma)$ we have:

$$\begin{cases} i = \frac{\lambda+\sigma}{2} - \alpha_0; \\ j = \frac{\sigma-\lambda}{2} - \beta_0. \end{cases} \qquad (27)$$

Then the following theorem can be formulated:

Theorem 5. *The mapping $\lambda(w)$ is an invariant of the reduction, i.e., $\lambda(w) = \lambda\left(w_{nf}\right)$, and the mapping $\sigma(w)$ is an invariant modulo $2p$ of the reduction, i.e., $\sigma(w) \equiv \sigma\left(w_{nf}\right) \mod (2p)$, where w_{nf} is the any word in $\mathcal{S}$ reduced to its normal form.*

The proof of this theorem follows from the definition of the reduction and thus we omit it.
The defined mappings have the following important property:

$$\lambda\left(w^k\right) = k\lambda(w); \qquad (28)$$

$$\sigma\left(w^k\right) = k\sigma(w). \qquad (29)$$

Let us assume that the entries of matrices Λ and Σ satisfy the conditions presented in Problem (21). Then the following one-to-one-mapping mapping can be defined:

$$\varphi_{(\alpha_0, \beta_0)}(\lambda, \sigma) = b^{\beta_0} a^i b^j a^{\alpha_0}, \qquad (30)$$

where the values of α_0 and β_0 are fixed.

Example 1. *Assume, that $\lambda = 3$ and $\sigma = 7$. Then we have:*

$$\varphi_{(0,0)}(3,7) = a^5 b^2;$$
$$\varphi_{(0,1)}(3,7) = a^4 b^2 a;$$
$$\varphi_{(1,0)}(3,7) = ba^5 b;$$
$$\varphi_{(1,1)}(3,7) = ba^4 ba.$$

Furthermore, if $\lambda = -3$ and $\sigma = 7$, then:

$$\varphi_{(0,0)}(-3,7) = a^2 b^5;$$
$$\varphi_{(0,1)}(-3,7) = ab^5 a;$$
$$\varphi_{(1,0)}(-3,7) = ba^2 b^4;$$
$$\varphi_{(1,1)}(-3,7) = bab^4 a.$$

However, $\varphi_{(\alpha_0, \beta_0)}(3,6)$ and $\varphi_{(\alpha_0, \beta_0)}(7,3)$ are undefined for any values of α_0 and β_0.

If we apply mapping $\varphi_{(\alpha_0, \beta_0)}$ to the pair of matrices (Λ, Σ) elementwise then we obtain a matrix $W = \{w_{ij}\}$, where the entries w_{ij} are defined as follows:

$$w_{ij} = \varphi_{(\alpha_0, \beta_0)}\left(\lambda_{ij}, \sigma_{ij}\right). \qquad (31)$$

Now we introduce the following expression:

$$X\left(\Lambda, \Sigma\right) Y = \left(X\Lambda Y, X\Sigma Y\right),$$

and apply the mapping $\varphi_{(\alpha_0, \beta_0)}$ to it. Due to Properties (28) and (29) we have:

$$\varphi_{(\alpha_0, \beta_0)}\left(X\Lambda Y, X\Sigma Y\right) = {}^X W^Y. \tag{32}$$

where the entries of matrix W are defined by Expression (31). Furthermore, we apply the mapping $\varphi_{(\alpha_0, \beta_0)}$ to the pair of matrices (B, C) in Problem (21) to obtain the following matrix:

$$\varphi_{(\alpha_0, \beta_0)}\left(B, C\right) = D,$$

where the entries of matrix D are defined by Expression (31). The two latter equations can be combined to yield $MPF_S^{\mathcal{N}^0, \mathcal{C}}$ problem, symbolically presented in Definition 1.

Theorem 6. *$MPF_S^{\mathcal{N}^0, \mathcal{C}}$ is NP-Complete.*

Proof. Due to the properties of mappings $\lambda(w)$ and $\sigma(w)$ in Expressions (25)–(27), the property of bijective mapping $\varphi_{(\alpha_0, \beta_0)}$ and Theorem 4, we find that CSBMMQ is a sub-problem of $MPF_S^{\mathcal{N}^0, \mathcal{C}}$. Since, according to Theorem 2, CSBMMQ is NP-Complete, then the $MPF_S^{\mathcal{N}^0, \mathcal{C}}$ problem is NP-Complete as well. □

Remark 10. *In fact, circulant MPF problem is NP-Complete in more general case, since for matrices X and Y with no zero entries only the upper left corner and bottom right corner entries of the base matrix W play an important role. More precisely the first and the last literal of the specified entries produce fixed values α_0 and β_0. Normal forms of other entries of the base matrix W are irrelevant.*

5. Conclusions

1. The proof of NP-Completeness of author's constructed MPF in previous Symmetry journal publication is presented. It is a new evidence, that this type of MPF can be considered for construction of a non-commuting cryptography primitive as a conjectured OWF.
2. The proof is based on two main approaches: we prove that certain GSAT is NP-Complete using modified Schaefer criteria, and, using this result, we prove that this GSAT is a sub-problem of the considered MPF problem. Hence this type of MPF problem is NP-Complete.
3. It is a new step to prove that KAP presented in our previous publication mentioned above has a provable security property.

Author Contributions: This article was supervised by E.S. who proposed the methology later improved by both authors. A.M. performed the investigation and analyzed the obtained results together with his supervisor. Both authors collected resources for the paper. A.M. wrote the paper.

Funding: This research received no external funding.

Conflicts of Interest: The authors declare no conflict of interest.

Abbreviations

The following abbreviations are used in this manuscript:

MPF	Matrix power function
OWF	one-way function
MQ problem	Multivariate quadratic problem
MMQ problem	Matrix MQ problem

BMMQ problem	Binary matrix MQ problem
SBMMQ problem	Singular binary matrix MQ problem
CSBMMQ problem	Constrained singular binary matrix MQ problem
GSAT problem	General satisfiability problem
NP-Commplete problem	Non-deterministic polynomial complete problem
CNF	Conjuntive normal form

References

1. Mihalkovich, A.; Sakalauskas, E. Asymmetric cipher based on MPF and its security parameters evaluation. In Proceedings of the Lithuanian Mathematical Society, Klaipeda, Lithuania, 11–12 June 2012; VU Matematikos ir Informatikos Institutas: Vilnius, Lithuania, 2012; Ser. A, Volume 53, pp. 72–77.
2. Mihalkovich, A.; Sakalauskas, E.; Venckauskas, A. New asymmetric cipher based on matrix power function and its implementation in microprocessors efficiency investigation. *Elektron. Elektrotech.* **2013**, *19*, 119–122. [CrossRef]
3. Sakalauskas, E.; Listopadskis, N.; Tvarijonas, P. Key Agreement Protocol (KAP) Based on Matrix Power Function. In *Advanced Studies in Software and Knowledge Engineering*; International Book Series "Information Science and Computing"; World Scientific: Singapore, 2008; pp. 92–96.
4. Sakalauskas, E.; Luksys, K. Matrix Power S-Box Construction. IACR Cryptology ePrint Archive 2007. Available online: http://citeseerx.ist.psu.edu/viewdoc/download?doi=10.1.1.78.2327&rep=rep1&type=pdf (accessed on 26 October 2018).
5. Sakalauskas, E.; Mihalkovich, A. New asymmetric cipher of non-commuting cryptography class based on matrix power function. *Informatica* **2014**, *25*, 283–298. [CrossRef]
6. Sakalauskas, E. The multivariate quadratic power problem over Zn is NP-Complete. *Inf. Technol. Control* **2012**, *41*, 33–39. [CrossRef]
7. Sakalauskas, E.; Mihalkovich, A. Improved Asymmetric Cipher Based on Matrix Power Function Resistant to Linear Algebra Attack. *Informatica* **2017**, *28*, 517–524. [CrossRef]
8. Sakalauskas, E.; Mihalkovich, A.; Venčkauskas, A. Improved asymmetric cipher based on matrix power function with provable security. *Symmetry* **2017**, *9*, 9. [CrossRef]
9. Shor, P.W. Polynomial-time algorithms for prime factorization and discrete logarithms on a quantum computer. *SIAM Rev.* **1999**, *41*, 303–332. [CrossRef]
10. Sakalauskas, E. Enhanced Matrix Power Function for Cryptographic Primitive Construction. *Symmetry* **2018**, *10*, 43. [CrossRef]
11. Garey, M.R.; Johnson, D.S. *Computers and Intractability*; WH Freeman: New York, NY, USA, 2002.
12. Patarin, J.; Goubin, L. Trapdoor one-way permutations and multivariate polynomials. In Proceedings of the International Conference on Information and Communications Security, Beijing, China, 11–14 November 1997; Springer: Berlin, Germany, 1997; pp. 356–368.
13. Schaefer, T.J. The complexity of satisfiability problems. In Proceedings of the Tenth Annual ACM Symposium on Theory of Computing, San Diego, CA, USA, 1–3 May 1978; ACM: New York, NY, USA, 1978; pp. 216–226.
14. Davis, P.J. *Circulant Matrices*; Wiley: New York, NY, USA, 1970.
15. Sakalauskas, E.; Mihalkovich, A. Candidate One-Way Function Based on Matrix Power Function with Conjugation Constraints. In Proceedings of the Conference proceedings Bulgarian Cryptography Days 2012, Sofia, Bulgaria, 20–21 September 2012; pp. 29–37.
16. Liu, J.; Zhang, H.; Jia, J. A linear algebra attack on the non-commuting cryptography class based on matrix power function. In Proceedings of the International Conference on Information Security and Cryptology, Beijing, China, 4–6 November 2016; Springer: Berlin, Germany, 2016; pp. 343–354.
17. Chrislock, J.L. On medial semigroups. *J. Algebra* **1969**, *12*, 1–9. [CrossRef]
18. Dechter, R.; Pearl, J. Structure identification in relational data. *Artif. Intell.* **1992**, *58*, 237–270. [CrossRef]

© 2018 by the authors. Licensee MDPI, Basel, Switzerland. This article is an open access article distributed under the terms and conditions of the Creative Commons Attribution (CC BY) license (http://creativecommons.org/licenses/by/4.0/).

Article

The Cracking of *WalnutDSA*: A Survey

José Ignacio Escribano Pablos [1,2], María Isabel González Vasco [1], Misael Enrique Marriaga [1] and Ángel Luis Pérez del Pozo [1,*]

1. MACIMTE, U. Rey Juan Carlos, 28933 Móstoles, Spain
2. BBVA Next Technologies, 28050 Madrid, Spain
* Correspondence: angel.perez@urjc.es

Received: 8 July 2019; Accepted: 20 August 2019; Published: 23 August 2019

Abstract: This paper reports on the Walnut Digital Signature Algorithm (*WalnutDSA*), which is an asymmetric signature scheme recently presented for standardization at the NIST call for post-quantum cryptographic constructions. *WalnutDSA* is a group theoretical construction, the security of which relies on the hardness of certain problems related to an action of a braid group on a finite set. In spite of originally resisting the typical attacks succeeding against this kind of construction, soon different loopholes were identified rendering the proposal insecure (and finally, resulting in it being excluded from Round 2 of the NIST competition). Some of these attacks are related to the well-structured and symmetric masking of certain secret elements during the signing process. We explain the design principles behind this proposal and survey the main attack strategies that have succeeded, contradicting its claimed security properties, as well as the recently-proposed ideas aimed at overcoming these issues.

Keywords: *WalnutDSA*; digital signatures; post-quantum cryptography; cryptanalysis

1. Introduction

The (seemingly close) advent of quantum computing is urging the cryptographic community to search for new constructions that may withstand attacks arising from this new computing paradigm. Post-quantum cryptography is a bursting research area in which tools are designed for a scenario where honest users are restricted to classical computation, while the adversary may eventually have access to quantum computing resources. The American National Institute of Standards and Technology (NIST) initiated in December 2016 *"a process to develop and standardize one or more additional public-key cryptographic algorithms [...] that are capable of protecting sensitive government information well into the foreseeable future, including after the advent of quantum computers"* (see [1]).

Walnut Digital Signature Algorithm (*WalnutDSA*) was one of the 20 public key signature schemes presented for standardization at the recent NIST call for post-quantum cryptographic constructions. Different mathematical objects were used in these proposals such as lattice theory, coding theory, algebraic geometry (see for instance [2–4]), and, in the case of *WalnutDSA*, braid groups. After a first round of evaluations, only nine of these proposals remained under consideration. *WalnutDSA* failed to enter the second round, mostly due to a number of attacks that were reported during the one-year evaluation phase.

While it is not unusual that post-quantum cryptographic proposals lack a formal security evaluation within the theoretical framework known as *provable security*, the lack of a rigorous security analysis of *WalnutDSA* has been particularly damaging for the scheme's credit. In particular, it makes it difficult to identify the critical points to fend off in an implementation. As a result, ad-hoc fixes have been proposed by the scheme designers after each published attack. Nevertheless, the effectiveness of these fixes is somewhat hard to judge. Moreover, the actual hardness of the underlying mathematical problems is not well understood. The signature process is simple and symmetric, having two secret group elements acting on the encoded message to be signed. However, this simplicity has been exploited in many of the attacks against the scheme. Unfortunately, several computational problems defined over the main algebraic environment of *WalnutDSA* (i.e., braid groups) have turned out to be hard to exploit cryptographically, mainly because the computational complexity of such problems may be high in a worst-case definition, while it is unclear how to produce hard instances effectively.

In this document, we give a self-contained review of *WalnutDSA*, detailing the proposal and describing the main attacks that have been presented against this construction, as well as the possible fixes, currently under discussion, towards a secure implementation of this signature scheme.

Paper roadmap: We start with a short section reviewing the history of braid group cryptography, followed by a section explaining the basics on signature schemes. Then, we give a comprehensive description of *WalnutDSA* in Section 4. Section 5 is devoted to the various attack strategies deployed against *WalnutDSA*; from the early factoring attacks (see Section 5.1), to collision attacks (Section 5.2), attempts to undermine the (claimed) one-wayness of the underlying *E-multiplication* function (Section 5.3), and finally, the recent (and probably the most devastating) attack aiming at the recovery of an alternative secret key by solving a certain rewriting problem (see Section 5.4). The survey wraps up with a short conclusion section.

2. Braid Group Cryptography

Cryptography based on braid groups was born almost 20 years ago and attracted plenty of attention from group theorists, as well as the cryptographic community. The reasons for this are diverse: the schemes were mathematically appealing and the constructions likely to be efficient enough to be practical. Unfortunately, many problems were brought to light after a thorough scrutiny carried out by pure mathematicians and cryptographers. In this section, we briefly review two of the most prominent proposals within this area and refer the interested reader to the survey on the topic by David Garber [5].

2.1. Cryptographic Constructions Using Braid Groups

The two flagship proposals made for deriving cryptographic constructions using braid groups are a key exchange protocol and a public key encryption scheme.

In 1999, Anshel, Anshel, and Goldfeld [6] introduced a generic two-party key establishment protocol. Their presentation could be translated into various implementations with different algebraic structures as a base (and, of course, security levels). The one using braid groups attracted the most attention. The security of this construction relied on the hardness of the so-called *multiple simultaneous conjugacy search problem* (see below) in the braid group.

Later, at CRYPTO 2000, Ko et al. [7] put forward a braid-based version of the Diffie–Hellman two-party key exchange protocol, as well as an encryption scheme á la ElGamal derived from such a protocol. The main idea behind this construction is as follows: Fix a public braid g. Using this public information and exchanging messages through a public channel, two users may establish a shared high entropy secret. This secret is derived from a braid of the form $(ab)g(ab)^{-1}$, which is constructed by letting each user choose a secret conjugating element (a and b respectively) and publicly interchanging the elements aga^{-1} and bgb^{-1}. Indeed, for this idea to work, the conjugating braids a and b should commute. Furthermore, the hardness of the underlying *conjugacy search problem* (see below) in the braid group is crucial for the security of the scheme, since extracting a or b from the public messages aga^{-1} and bgb^{-1} is enough to deduce the exchanged key.

2.2. Computational Problems in Braid Groups

Many cryptographic proposals (like the ones mentioned above) based their security in computational problems related to the so-called *conjugacy problem* in B_n, the braid group on $n > 0$ strands. However, assuming that these problems are hard is not always reasonable. Indeed, efficient algorithms for special cases of these problems have been behind the cryptanalysis of most of the cryptographic proposals designed using braid groups. Some examples of such problems are:

- *Conjugacy Decision Problem* (CDP). Given $A, B \in B_n$, determine whether they are conjugate, i.e., whether there exists $X \in B_n$ such that $A = X^{-1}BX$.
- *Conjugacy Search Problem* (CSP). Given $A, B \in B_n$, known to be conjugate, compute $X \in B_n$ such that $A = X^{-1}BX$.
- *Braid Diffie–Hellman Decision Problem* (BDHDP). Given $A, B, C, D \in B_n$, such that there exist $X, Y \in B_n$ satisfying $B = X^{-1}AX$ and $C = Y^{-1}AY$, with $XY = YX$, determine whether $D = Y^{-1}X^{-1}AXY$.
- *Braid Diffie–Hellman Search Problem* (BDHSP). Given $A, B, C \in B_n$, such that there exist $X, Y \in B_n$ satisfying $B = X^{-1}AX$ and $C = Y^{-1}AY$, with $XY = YX$, compute $D = Y^{-1}X^{-1}AXY$.
- *Multiple Simultaneous Conjugacy Search Problem* (MSCSP). Given k pairs of elements $(U_i, W_i) \in B_n^2$, such that they are all conjugates with respect to the same braid, find such a conjugating braid, i.e., compute $X \in B_n$ such that $W_i = X^{-1}U_iX$, for all $i = 1, \ldots, k$.
- *Decomposition Problem* (DP). Let G be a fixed subgroup of B_n. Given $A, B \in B_n$, find $X, Y \in G$ such that $B = XAY$.
- *Root Extraction Problem* (REP). For $A \in B_n$ and $r \in \mathbb{N}$ such that there exists $B \in B_n$ with $A = B^r$, compute such a braid B.

It is easy to see that there are close relations among the above problems. Let us focus on how to solve CSP and CDP. As explained in detail in [5], the basic idea that has proven more fruitful towards a solution for the CSP and CDP problems involves a set I_x for each braid x (typically a subset of the conjugacy class of A), which characterizes the conjugacy class (i.e., A and B are conjugates if and only if $I_A = I_B$). Furthermore, there should be an efficient algorithm to compute a representative $\hat{A} \in I_A$ and a *witness* $X \in B_n$, such that $X^{-1}AX = \hat{A}$. Last, it should be possible to construct the full set I_A in a finite number of steps, starting from any representative $\hat{A}$. Now, given two braids $A, B \in B_n$, specifying an instance of CSP or CDP, one should:

(i) find representatives $\hat{A} \in I_A$ and $\hat{B} \in I_B$;
(ii) compute elements of I_A (storing the corresponding witnesses) until either:

  (a) $\hat{B}$ is found as an element of I_A, proving A and B to be conjugate and providing a conjugating element or
  (b) the entire set I_A is constructed without finding $\hat{B}$, proving that A and B are not conjugate.

Several choices of the special sets I_A can be found in the literature: summit sets, super summit sets, ultra summit sets, reduced supper summit sets, etc. All of them are subsets of the conjugacy class of the corresponding braid A. Of course, choosing a simpler and smaller set results in a more efficient algorithm derived from the above strategy. Using the above technique and other sophisticated geometric techniques, Birman, Gebhardt, and González Meneses [8] provided a polynomial-time algorithm to solve the CSP involving the so-called *periodic braids*. Furthermore, the same authors proved that the problem would be solved for all instances if a polynomial-time algorithm for a special type of braid (*rigid braids*) was found.

However, not only full theoretical solutions for the conjugacy problems have been of interest in the cryptographic context; indeed, heuristic algorithms with a significant success rate suffice to thwart the security of a scheme that is based on one of the above problems (we refer again to [5] for details). As a consequence, all cryptographic proposals built around the above problems are currently considered problematic.

3. Basics on Signature Schemes

In this section, we recall some basic concepts related to public key digital signature schemes and the assessment of provable security for these cryptographic tools. Many of the definitions below are taken from [9,10].

Definition 1. *A digital signature scheme is a triplet of algorithms* $(\mathcal{G}, \Sigma, \mathcal{V})$ *where:*

- $\mathcal{G}$, *the key generation algorithm, is a probabilistic algorithm that takes as input* $\mathbf{1}^\lambda$ *(for a security parameter* $\lambda \in \mathbb{N}$) *and returns a pair* $(\mathbf{pk}, \mathbf{sk})$ *of public and secret keys, from a designated key space of polynomial size in* λ.
- Σ, *the signing algorithm, is a probabilistic algorithm that takes as input a given message* $m \in \mathcal{M}_\lambda$ *(for a fixed message space) and a secret key* $\mathbf{sk}$ *and returns a signature* sig *(also assumed to belong to a prescribed set of polynomial size in* λ). *ss of generality, we can assume that each* $\mathcal{M}_\lambda$ *consists of bitstrings of polynomial size in* λ. *In the sequel, we often drop the subscript* λ *for the sake of readability*
- $\mathcal{V}$, *the verification algorithm, is a deterministic algorithm that takes as input a given signature* sig, *a message* $m \in \mathcal{M}_\lambda$, *and a public key* $\mathbf{pk}$ *and outputs a bit in* $\{0, 1\}$, *checking if* sig *is a valid signature of m with respect to* $\mathbf{pk}$.

Typically, a correctness requirement is imposed, establishing that $\mathcal{V}$ outputs one if it gets a valid signature as the input. The fact that it should output zero for an invalid signature is typically captured by the different definitions of security.

3.1. Security Notions for Signature Schemes

Prior to giving formal definitions of security notions, we informally list the different adversarial goals and attack models, which attempt to capture the main attack strategies that should be prevented for each specific adversary. Let $\mathcal{A}$ denote a (probabilistic polynomial-time) adversary. We assume that $\mathcal{A}$ pursues one of the following *adversarial goals*:

- Existential Forgery (EF): $\mathcal{A}$ tries to produce a valid signature for a message m, not necessarily adversarial chosen.
- Selective Forgery (SF): $\mathcal{A}$ tries to produce a valid signature for some adversarial chosen fixed message m.
- Universal Forgery (UF): $\mathcal{A}$ aims at producing a valid signature for *any* given message.
- Total Break (TB): $\mathcal{A}$ tries to retrieve, from the public information, a legitimate signer's secret key.

Similarly, in order to capture adversarial capabilities, we distinguish among the following *attack models*:

- No Message Attack (NMA): $\mathcal{A}$ only knows the public parameters (in particular, the public signing key).
- Random Message Attack (RMA): $\mathcal{A}$ is given signatures on a sequence of messages selected uniformly at random.
- Chosen Message Attack (CMA): $\mathcal{A}$ is given access to a signing oracle, which signs any message chosen by $\mathcal{A}$. Queries to this oracle can be adaptive, i.e., $\mathcal{A}$ may adapt the input messages based on previous output signatures.

Formal security notions are introduced by combining adversarial goals and capabilities. For instance, a signature scheme is secure in the sense of UF-NMA if given any probabilistic polynomial-time adversary $\mathcal{A}$, there exists a negligible function of the security parameter bounding the probability of success of a UF attack, provided that $\mathcal{A}$ has access only to public information (NMA). Other security notions are defined analogously; for instance, EUF-CMA captures the fact that a CMA adversary will not be able to produce an existential forgery.

Now, we give precise definitions for the three security notions, which are relevant throughout this work.

Definition 2. *A signature scheme* $(\mathcal{G}, \Sigma, \mathcal{V})$ *with message space* $\mathcal{M}$ *and security parameter* λ *is said to be universally unforgeable under no-message attacks (UF-NMA) if for any probabilistic polynomial-time adversary* $\mathcal{A}$ *and* $\forall m \in \mathcal{M}$*, then:*

$$\Pr \begin{bmatrix} (\mathbf{pk}, \mathbf{sk}) \leftarrow \mathcal{G}(\mathbf{1}^\lambda); \\ \mathtt{sig} \leftarrow \mathcal{A}(\mathbf{pk}, m); \\ \mathcal{V}(m, \mathtt{sig}, \mathbf{pk}) = 1 \end{bmatrix} \leq \mathrm{negl}(\lambda).$$

Definition 3. *A signature scheme* $(\mathcal{G}, \Sigma, \mathcal{V})$ *with message space* $\mathcal{M}$ *and security parameter* λ *is said to be universally unforgeable under random-message attacks (UF-RMA) if for any probabilistic polynomial-time adversary* $\mathcal{A}$ *and* $\forall m \in \mathcal{M}$*, then:*

$$\Pr \begin{bmatrix} (\mathbf{pk}, \mathbf{sk}) \leftarrow \mathcal{G}(\mathbf{1}^\lambda); \\ \{m_i\}_{i=1}^k \xleftarrow{\$} \mathcal{M} \setminus \{m\}; \\ \{\mathtt{sig}_i\}_{i=1}^k \leftarrow \Sigma(\mathbf{sk}, \{m_i\}_{i=1}^k); \\ \mathtt{sig} \leftarrow \mathcal{A}(\mathbf{pk}, \{(m_i, \mathtt{sig}_i)\}_{i=1}^k, m); \\ \mathcal{V}(m, \mathtt{sig}, \mathbf{pk}) = 1 \end{bmatrix} \leq \mathrm{negl}(\lambda).$$

The above definition states that when given a list of message-signature pairs, where the messages are selected uniformly at random, the adversary should still have only a negligible probability of constructing a new valid signature pair.

Definition 4. *A signature scheme* $(\mathcal{G}, \Sigma, \mathcal{V})$ *with message space* $\mathcal{M}$ *and security parameter* λ *is said to be existentially unforgeable under adaptive chosen-message attacks (EUF-CMA) if for any probabilistic polynomial-time adversary* $\mathcal{A}$ *with polynomial access to a signing oracle* $\mathcal{O}_{\mathbf{sk}}$ *that produces valid signatures with respect to a certain secret key* $\mathbf{sk}$*, then:*

$$\Pr \begin{bmatrix} (\mathbf{pk}, \mathbf{sk}) \leftarrow \mathcal{G}(\mathbf{1}^\lambda); \\ \{\mathtt{sig}_i\}_{i=1}^k \leftarrow \mathcal{O}_{\mathbf{sk}}(\{m_i\}_{i=1}^k); \\ (m, \mathtt{sig}) \leftarrow \mathcal{A}(\mathbf{pk}, \{(m_i, \mathtt{sig}_i)\}_{i=1}^k); \\ \mathcal{V}(m, \mathtt{sig}, \mathbf{pk}) = 1 \text{ and } m \notin \{m_1, \ldots, m_k\} \end{bmatrix} \leq \mathrm{negl}(\lambda).$$

In the above definition, the adversary is given access to a signing oracle that produces valid signatures with respect to the key pair under attack and faces the challenge of producing a valid signature for a message. This model is particularly relevant for capturing malleability attacks, which exploit the possibility of deriving new valid signatures from legitimate ones.

The standard security definition for signature schemes is EUF-CMA, which is the strongest among the three notions we have introduced. More precisely, every EUF-CMA scheme is UF-RMA, and in turn, every UF-RMA scheme is UF-NMA.

4. Scheme Description

In this section, we describe the *Walnut Digital Signature Algorithm (WalnutDSA)* introduced in [11]. This construction relies on certain computational properties of nonlinear operations in the *Artin braid group* B_N [12] combined with operations in $\mathrm{GL}_N(\mathbb{F}_q)$, the group of non-singular $N \times N$ matrices with entries in the finite field $\mathbb{F}_q$ with q elements.

Informally, in *WalnutDSA*, the message to be signed is hashed and encoded as a braid in B_N, (see Section 4.1). The private key consists of a pair of braids, while an ordered set of N elements in $\mathbb{F}_q$ and a pair of elements of the set $\mathrm{GL}_N(\mathbb{F}_q) \times S_N$ form the public key (As usual, S_N is the group of permutations of $\{1, 2, \ldots, N\}$). Key generation is described in detail in Section 4.2. In order to render brute force attacks ineffective, the key space is made sufficiently large by choosing $N \geq 8$ and $q \geq 32$. A signature is built from the encoded message, the private keys, and two additional braids used to obscure the private key. Valid signatures must verify a certain equation involving the public key, the encoded message, and *E-multiplication*, a group-theoretic one-way function introduced in [13]. All these algorithms are precisely described in Section 4.3. Let us start here by describing the mathematical ingredients needed to understand them.

4.1. Message Encoding

WalnutDSA encodes messages as elements in the Artin braid group, which is a nice algebraic and computational habitat.

4.1.1. Braids

Informally, the *braid group with N strands* B_N is a non-Abelian group whose elements can be described as a configuration of N non-intersecting vertical or horizontal strands in three-dimensional space, with ends fixed on two parallel disks. Moreover, the strands flow in one direction without turning back, so that any plane parallel to the disks will intersect each strand exactly once. Multiplication of two braids is defined as concatenation of strands, and two braids are considered equal if one can be continuously transformed into the other, keeping the ends fixed and without intersecting the strands.

More precisely, the braid group with N strands is defined as follows [12]. For $N \geq 2$, B_N is a group generated by the *Artin generators* $\{b_1, b_2, \ldots, b_{N-1}\}$, subject to the following relations:

$$\begin{aligned} b_i\, b_{i+1}\, b_i &= b_{i+1}\, b_i\, b_{i+1}, & 1 \leq i \leq N-2, \\ b_i\, b_j &= b_j\, b_i, & |i-j| \geq 2. \end{aligned} \quad (1)$$

The Artin generator b_i represents the braid where the ith strand crosses over the $(i+1)$th strand. The relation $b_i\, b_{i+1}\, b_i = b_{i+1}\, b_i\, b_{i+1}$ for $1 \leq i \leq N-2$, corresponds to moving the ith strand over the crossing of the $(i+1)$th and the $(i+2)$th strand, and the relations $b_i\, b_j = b_j\, b_i$ for $|i-j| \geq 2$ correspond to the fact that crossings that do not share strands commute.

Any braid $b \in B_N$ can be expressed as a product of the Artin generators and their inverses, that is,

$$b = b_{i_1}^{e_1}\, b_{i_2}^{e_2} \cdots b_{i_k}^{e_k}, \quad (2)$$

where $1 \leq i_n \leq N-1$ and $e_n \in \{-1, 1\}$. Clearly, the expression for b is not unique since applying (1) yields infinite equivalent expressions.

Let S_N be the symmetric group of order N. There exists a group homomorphism $\sigma : B_N \to S_N$ defined as follows. For each Artin generator b_i, $\sigma_{b_i}(i) = \sigma_{b_i^{-1}}(i) = i+1$ and $\sigma_{b_i}(j) = \sigma_{b_i^{-1}}(j) = j$ for $1 \leq j \leq N-1$, such that $i \neq j$. That is, b_i and b_i^{-1} are mapped into the element in S_N, which interchanges the ith and the $(i+1)^{\text{th}}$ elements of $\{1, 2, \ldots, N\}$ and leaves the rest fixed. Notice that $\sigma_{b_i}^{-1} = \sigma_{b_i}$ for $1 \leq i \leq N-1$. Moreover, for $1 \leq i, j \leq N-1$, $\sigma_{b_i^{\pm 1}\, b_j^{\pm 1}} = \sigma_{b_i}\, \sigma_{b_j}$. Hence, for any braid $b \in B_N$ as in (2), we have:

$$\sigma_b = \sigma_{b_{i_1}}\, \sigma_{b_{i_2}} \cdots \sigma_{b_{i_k}}.$$

If σ_b is the identity element of S_N, then b is called a *pure braid*. In other words, a braid is a pure braid if and only if it is in the kernel of σ.

4.1.2. Encoding

WalnutDSA requires the permutation linked to each encoded message to be the identity. Thus, the encoded message must be a pure braid.

The encoding algorithm utilizes the following collection of pure braids:

$$g_{N,i} = b_{N-1} b_{N-2} \cdots b_{i+1} b_i^2 b_{i+1}^{-1} \cdots b_{N-2}^{-1} b_{N-1}^{-1}, \quad 1 \leq i \leq N-1.$$

This collection of pure braids generates a free subgroup of B_N [14], that is the set of products of $g_{N,i}$, $1 \leq i \leq N-1$, that satisfy no relations except those implied by the group axioms (e.g., $a\,b = a\,c\,c^{-1}\,b$, but $a \neq b^{-1}$ for $a, b, c \in B_N$) ([15], Chapter 7). Any subset of the above collection of pure braids will generate a free subgroup.

Let $m \in \{0,1\}^*$ be a message, and let $H : \{0,1\}^* \to \{0,1\}^{4\ell}$, $\ell \geq 1$, denote a cryptographically-secure hash function. Fix any four generators g_{N,j_1}, g_{N,j_2}, g_{N,j_3}, g_{N,j_4}, and denote by $C_{N,4}$ the free subgroup generated by these four generators. Define the encoding function $E : \{0,1\}^{4\ell} \to C_{N,4}$ as follows. The hashed message $H(m)$ is broken into ℓ 4-bit blocks. For the kth block, the first two bits determine a generator $g_{N,j_{n_k}}$, $1 \leq n_k \leq 4$, and the next two bits determine an integer $1 \leq p_k \leq 4$. Then,

$$E(H(m)) = g_{N,j_{n_1}}^{p_1} g_{N,j_{n_2}}^{p_2} \cdots g_{N,j_{n_\ell}}^{p_\ell},$$

written in its reduced form, that is products of the form $b_i b_i^{-1}$ and $b_i^{-1} b_i$, $1 \leq i \leq N-1$, are erased from the braid (see [16,17] for examples of reduction algorithms). This encoding algorithm ensures that each message is mapped to a unique reduced element of the free subgroup generated by g_{N,j_1}, g_{N,j_2}, g_{N,j_3}, g_{N,j_4}.

4.2. Key Generation

The security of *WalnutDSA* relies on E-multiplication, a function that maps braids in B_N to elements in the set $\mathrm{GL}_N(\mathbb{F}_q) \times S_N$. This mapping is based on the colored Burau representation of B_N. We provide some preliminaries before describing the public and private keys in *WalnutDSA*.

4.2.1. Colored Burau Representation of the Braid Groups

Let $\mathcal{L}_{\mathbb{F}_q} \equiv \mathcal{L}_{\mathbb{F}_q}[t_1, t_2, \ldots, t_N]$ denote the ring of Laurent polynomials in the variables $t_1, t_2, \ldots, t_N$ with coefficients in $\mathbb{F}_q$, that is,

$$\mathcal{L}_{\mathbb{F}_q} = \left\{ \sum_{j=0}^{k} a_j t_1^{n_{1,j}} t_2^{n_{2,j}} \cdots t_N^{n_{N,j}} : n_{i,j} \in \mathbb{Z}, a_j \in \mathbb{F}_q, k \geq 0 \right\}.$$

For each Artin generator, we define the following $N \times N$ matrices [18]:

$$\mathrm{CB}_{b_1}(t_1) = \left(\begin{array}{cc|c} -t_1 & 1 & 0 \\ 0 & 1 & 0 \\ \hline 0 & 0 & I_{N-2} \end{array} \right),$$

$$\mathrm{CB}_{b_1^{-1}}(t_2) = \left(\begin{array}{cc|c} -t_2^{-1} & t_2^{-1} & 0 \\ 0 & 1 & 0 \\ \hline 0 & 0 & I_{N-2} \end{array} \right),$$

$$\mathrm{CB}_{b_i}(t_i) = \left(\begin{array}{c|ccc|c} I_{i-2} & 0 & 0 & 0 & 0 \\ \hline 0 & 1 & 0 & 0 & 0 \\ 0 & t_i & -t_i & 1 & 0 \\ 0 & 0 & 0 & 1 & 0 \\ \hline 0 & 0 & 0 & 0 & I_{N-i-1} \end{array} \right), \quad 2 \leq i \leq N-1,$$

$$\mathrm{CB}_{b_i^{-1}}(t_{i+1}) = \left(\begin{array}{c|ccc|c} I_{i-2} & 0 & 0 & 0 & 0 \\ \hline 0 & 1 & 0 & 0 & 0 \\ 0 & 1 & -t_{i+1}^{-1} & t_{i+1}^{-1} & 0 \\ 0 & 0 & 0 & 1 & 0 \\ \hline 0 & 0 & 0 & 0 & I_{N-i-1} \end{array} \right), \quad 2 \leq i \leq N-1,$$

where I_n is the identity matrix of size $n \times n$ and 0 is the zero matrix of adequate size.

Let $M \in \text{GL}_N(\mathcal{L}_{\mathbb{F}_q})$ and $\pi \in S_N$. We define:

$$^\pi M(t_1, t_2, \ldots, t_N) = M(t_{\pi(1)}, t_{\pi(2)}, \ldots, t_{\pi(N)}).$$

The product of (A, π) and (B, τ) in $\text{GL}_N(\mathcal{L}_{\mathbb{F}_q}) \times S_N$ is defined as:

$$(A, \pi) \cdot (B, \tau) = (A \cdot {}^\pi B, \pi\tau).$$

We have that the elements of $\text{GL}_N(\mathcal{L}_{\mathbb{F}_q}) \times S_N$ form a group under this product operation.

Now, we define the colored Burau representation:

$$\Pi_{\text{CB}} : B_N \to \text{GL}_N(\mathcal{L}_{\mathbb{F}_q}) \times S_N.$$

For any Artin generator b_i, $1 \leq i \leq N-1$,

$$\Pi_{\text{CB}}(b_i^{\pm 1}) = (\text{CB}_{b_i^{\pm 1}}, \sigma_{b_i}),$$

and for all b_i and b_j, $1 \leq i, j \leq N-1$,

$$\Pi_{\text{CB}}(b_i^{\pm 1} b_j^{\pm 1}) = \Pi_{\text{CB}}(b_i^{\pm 1}) \cdot \Pi_{\text{CB}}(b_j^{\pm 1}).$$

More generally, for any braid $b \in B_N$ as in (2),

$$\Pi_{\text{CB}}(b) = \Pi_{\text{CB}}(b_{i_1}^{e_1}) \cdot \Pi_{\text{CB}}(b_{i_2}^{e_2}) \cdots \Pi_{\text{CB}}(b_{i_k}^{e_k}).$$

It can be verified that Π_{CB} is a homomorphism that preserves the braid relations (1) and, hence, defines a representation of B_N.

4.2.2. E-Multiplication

The key generation in *WalnutDSA* is based on E-multiplication, a group-theoretic one-way function introduced in [13]. Here, we recall its definition.

Fix a finite field $\mathbb{F}_q$ and a set of N non-zero elements in $\mathbb{F}_q$,

$$\mathcal{T} = \{y_1, y_2, \ldots, y_N\} \subset \mathbb{F}_q.$$

For every $M \in \text{GL}_N(\mathcal{L}_{\mathbb{F}_q})$, we define:

$$M \downarrow_\mathcal{T} = M(y_1, y_2, \ldots, y_N) \in \text{GL}_N(\mathbb{F}_q).$$

Now, we define E-multiplication.

Definition 5. *E-multiplication, denoted by $\star$, is a right action of the group $\Pi_{\text{CB}}(B_N)$ on the set $\text{GL}_N(\mathbb{F}_q) \times S_N$, defined inductively as follows. Given $(M, \pi) \in \text{GL}_N(\mathbb{F}_q) \times S_N$,*

$$(M, \pi) \star \Pi_{\text{CB}}(b_i^{\pm 1}) = (M \cdot {}^\pi \text{CB}_{b_i^{\pm 1}} \downarrow_\mathcal{T}, \pi\sigma_{b_i}) \in \text{GL}_N(\mathbb{F}_q) \times S_N, \qquad 1 \leq i \leq N-1.$$

More generally, for any braid $b \in B_N$ as in (2),

$$(M, \pi) \star \Pi_{\text{CB}}(b) = (M, \pi) \star \Pi_{\text{CB}}(b_{i_1}^{e_1}) \star \Pi_{\text{CB}}(b_{i_2}^{e_2}) \star \cdots \star \Pi_{\text{CB}}(b_{i_k}^{e_k}),$$

where the operations are done from left to right. Furthermore, for convenience, we will write $(M, \pi) \star b$ instead of $(M, \pi) \star \Pi_{\text{CB}}(b)$.

4.2.3. Key Generation Mechanism

The signer's private key consists of two random braids s_1 and s_2, written in reduced form, such that s_1, s_2, and $s_1 s_2$ are not pure braids. No further prerequisites were made explicit in the original proposal of *WalnutDSA*.

Let $b \in B_N$ be a braid and $\mathcal{T} \subset \mathbb{F}_q$ a fixed set of N non-zero elements. Define:

$$\mathcal{P}(b) \equiv \mathcal{P}_{\mathcal{T}}(b) = (I_N, \iota_N) \star b \in \mathrm{GL}_N(\mathbb{F}_q) \times S_N,$$

where I_N is the identity $N \times N$ matrix and $\iota_N \in S_N$ is the identity permutation. The signer's public key consists of:

- $\mathcal{T} = \{y_1, y_2, \ldots, y_N\} \subset \mathbb{F}_q$ such that $y_i \neq 0$, $1 \leq i \leq N$, and $y_a = y_b = 1$ for some $1 \leq a, b \leq N$,
- $\mathcal{P}(s_1)$, and
- the matrix component of $\mathcal{P}(s_2)$, denoted by $\mathtt{mat}(\mathcal{P}(s_2))$, that is, $\mathtt{mat}(\mathcal{P}(s_2))$.

4.3. Signature Generation and Verification

We now describe *WalnutDSA* in detail.

4.3.1. Cloaking Elements

First, we discuss elements in the subgroup of pure braids that essentially disappear when performing E-multiplication. The purpose of these elements is to *cloak*, or hide, the private key used to construct the signature.

Definition 6. *Let $(M, \pi) \in \mathrm{GL}_N(\mathbb{F}_q) \times S_N$, and let $\mathcal{T}$ be a fixed set of N non-zero elements of $\mathbb{F}_q$. A pure braid $v \in B_N$ is called a cloaking element of (M, π) if:*

$$(M, \pi) \star v = (M, \pi).$$

It is clear from this definition that the set of cloaking elements of (M, π) depends on the set $\mathcal{T}$. The existence of cloaking elements is discussed in the following proposition.

Proposition 1. *Fix integers $N \geq 2$, $1 \leq a, b \leq N$, and fix a set of N non-zero elements $\mathcal{T} = \{y_1, y_2, \ldots, y_N\} \subset \mathbb{F}_q$ such that $y_a = y_b = 1$. Let $(M, \pi) \in \mathrm{GL}_N(\mathbb{F}_q) \times S_N$, b_i, $1 \leq i \leq N-1$, an Artin generator of B_N, and $w \in B_N$ such that:*

$$\pi \sigma_w(i) = a, \quad \text{and} \quad \pi \sigma_w(i+1) = b.$$

Then,

$$v = w b_i^2 w^{-1},$$

is a cloaking element of (M, π).

4.3.2. Signature Generation

Fix a hash function $H : \{0,1\}^* \to \{0,1\}^{4\ell}$, $\ell \geq 1$, and let $s_1, s_2 \in B_N$ be the braids in the private key. In *WalnutDSA*, a signature for the message $m \in \{0,1\}^*$ is the braid:

$$\mathtt{sig} = v_1 \cdot s_1^{-1} \cdot v \cdot E(H(m)) \cdot s_2 \cdot v_2 \in B_N,$$

written in reduced form, where $v, v_1, v_2 \in B_N$ are cloaking elements of (I_N, ι_N), $\mathcal{P}(s_1)$, and $\mathcal{P}(s_2)$, respectively.

4.3.3. Signature Verification

The verification algorithm calculates the matrix component of $\mathcal{P}(s_1) \star \text{sig}$ and $\mathcal{P}(E(H(m)))$, denoted by $\text{mat}(\mathcal{P}(s_1) \star \text{sig})$ and $\text{mat}(\mathcal{P}(E(H(m))))$, respectively, and accepts the signature if the following equation holds:

$$\text{mat}(\mathcal{P}(s_1) \star \text{sig}) = \text{mat}(\mathcal{P}(E(H(m)))) \cdot \text{mat}(\mathcal{P}(s_2)).$$

5. Cryptanalysis of *WalnutDSA*

5.1. Factoring Attacks

The essential idea behind these attacks is to forge a signature for any given message m solving a factorization problem in groups, defined as follows:

Definition 7 (Factorization problem in groups). *Let G be a group; let $\Gamma = \{g_1, \ldots, g_\gamma\}$ be a generating set for G; and let $h \in G$. Find an integer L and sequences $(k_1, \ldots, k_L) \in \{1, \ldots, \gamma\}^L$ and $(\varepsilon_1, \ldots, \varepsilon_L) \in \{\pm 1\}^L$ such that:*

$$h = \prod_{i=1}^{L} g_{k_i}^{\varepsilon_i}.$$

A solution to a specific instance of this problem has been exploited by several authors [19,20] to construct a new valid signature from several valid signatures, in order to violate UF-CMA. More precisely, Hart et al. presented in [19] an efficient method to compute, given a couple of signatures on random messages, a new signature on an arbitrary message. However, these forged signatures were significantly longer than those constructed by the honest signer. The design of *WalnutDSA* was modified by the authors in order to defeat this attack, yet a refinement of this method, presented in Section 3 of [20], rendered this modification insufficient.

5.1.1. Factoring For Universal Forgeries: The Attacks by Hart Et Al., and Beullens and Blackburn

The strategy behind [19] allows for constructing a valid signature for any arbitrary message m (and is thus a universal forgery). More precisely, Proposition 4 in [19] states that, given a finite set of signatures:

$$\mathcal{S} = \{(m_i, \text{sig}_i) : i \in I\}$$

and taking g_i as the matrix part of $\mathcal{P}(E(H(m_i)))$ for all $i \in I$, it holds that, if the matrix part h of $\mathcal{P}(E(H(m)))$ can be factored with respect to the generating set $\{g_i \mid i \in I\}$, then constructing the very same *word* replacing each g_i with the corresponding braids s_i from $\mathcal{S}$ yields a valid signature for m.

Beullens and Blackburn explained how to exploit this malleability property through the following simple theorem:

Theorem 1 (Theorem 1 from [20]). *Consider the version of WalnutDSA, where it holds that $s_1 = s_2$. Suppose m, m_1, m_2 are three messages. Let h, h_1, h_2 be the matrix part of $\mathcal{P}(E(H(m)))$, $\mathcal{P}(E(H(m_1)))$, $\mathcal{P}(E(m_2))$, respectively. Then,*

1. *If $h = h_1^{-1}$ and sig_1 is a valid signature for m_1, then sig_1^{-1} is a valid signature for m.*
2. *If $h = h_1 \cdot h_2$ and sig_1, sig_2 are valid signatures for m_1 and m_2, respectively, then $\text{sig}_1 \cdot \text{sig}_2$ is a valid signature for m.*

However, the above result is only valid if the public braids s_1 and s_2 coincide, which was only the case in the first versions of the proposal [11]. All in all, a simple variant of the above theorem, presented in [20], evidences that choosing $s_1 \neq s_2$ does not amend the strong malleability inherent to *WalnutDSA*:

Theorem 2 ([20])**.** *Suppose m, m_1, m_2 are three messages. Let h, h_1, h_2 be the matrix part of $\mathcal{P}(E(H(m)))$, $\mathcal{P}(E(H(m_1)))$, $\mathcal{P}(E(H(m_2)))$, respectively. Let $s_1, s_2, s_3 \in B_N$ be three braids. Then,*

1. *If $h = h_1^{-1}$ and sig_1 is a valid signature for m_1 under the public key $(\mathcal{P}(s_1), \mathcal{P}(s_2))$, then sig_1^{-1} is a valid signature for m under the public key $(\mathcal{P}(s_2), \mathcal{P}(s_1))$.*
2. *If $h = h_1 \cdot h_2$ and $\text{sig}_1 \cdot \text{sig}_2$ are valid signatures for m_1 and m_2 under the public keys $(\mathcal{P}(s_1), \mathcal{P}(s_2))$ and $(\mathcal{P}(s_2), \mathcal{P}(s_3))$, respectively, then $\text{sig}_1 \cdot \text{sig}_2$ is a valid signature for m under the public key $(\mathcal{P}(s_1), \mathcal{P}(s_3))$.*

Note that the above theorems do not impose a practical restriction on the forged message m, for suitable m_1, m_2 can be constructed for any m in order to mount the UF attack. Still, the forged signatures obtained through these factoring strategies are many orders of magnitude longer than legitimate signatures; thus, imposing length limits on the output signatures (as the authors did in the implementation submitted to the NIST PQCstandardization call) is enough to dodge these attacks.

5.1.2. Factoring Using the Garside Normal Form

Recently, in [21], it was noticed that whenever a product of braids $ABC \in B_N$ is represented in the Garside normal form, parts of the corresponding form of the individual factors A, B, and C are somewhat easy to extract. In particular, the authors of this paper presented an algorithm for recovering, given B, elements A' and C' such that:

- $A = A'$, $C = C'$ up to multiplications with elements in the center of B_N
- $AC = A'C'$.

Note that the center of the group B_N is a cyclic group generated by the square of the so-called *Garside's fundamental braid*, Δ, which is the only positive braid for which any two strands cross exactly once (see [14,22] for a classical introduction and a comprehensive survey on braid groups). This decomposition strategy allows for constructing a universal forgery, as stated in the following result:

Theorem 3 ([21])**.** *Let $W_1 \cdot E(H(m)) \cdot W_2 \in B_N$ be a valid signature for some message m, and let W_1', $W_2' \in B_N$ such that $W_1' \equiv W_1 \mod \Delta^2$, $W_2' \equiv W_2 \mod \Delta^2$, and $W_1 \cdot W_2 = W_1' \cdot W_2'$. Then,*

$$W_1' \cdot E(H(m')) \cdot W_2'$$

is a valid signature for any message m'.

Note that since the *replaced* braids W_1 and W_2 are in principle independent of the message m, the forged signature need not be longer than a legitimate signature. Furthermore, the complexity of this procedure is essentially that of computing Garside normal forms, which can be done in time $\mathcal{O}(k^2 N)$, where k is the number of Artin generators encoding the input braid.

Furthermore, this method fends off the colored Burau representation used in the implementation of *WalnutDSA*; thus, it cannot be prevented by modifying the size of the underlying finite field. The authors of this cryptanalysis suggest that the only way to dodge this attack is to add many concealed cloaking elements to the encoding, which has a significant cost both in signature length and computing time for the generation of signatures. Furthermore, in [23], the authors of the scheme claimed to have experimentally demonstrated that inserting cloaking elements every 7–12 generators into the braid $E(H(m'))$ blocked this attack. However, no details were given on how this strategy was theoretically or empirically assessed.

5.2. Collision Attacks

Imposing implicit limits on the output signature sizes is indeed a valid strategy for preventing factoring attacks, and so, it was promptly noticed by the authors of *WalnutDSA*. However, in Section 4 of [20], it was demonstrated that, through a simple collision method, it was possible to compute *short* forged signatures, yet not on arbitrary messages.

In Section 4 of [20], it was observed that if there exist two messages m_1, m_2 such that $\mathcal{P}(E(H(m_1))) = \mathcal{P}(E(H(m_2)))$, then a valid signature for m_1 is valid for m_2 and vice versa. Breaking the EUF-CMA security notion (see Definition 4) is as simple as finding such two messages m_1 and m_2, since an adversary could query a signature for m_1 and then obtain a signature for m_2.

A generic collision attack is expected to require $|\mathcal{P}(E(\{0,1\}^*))|^{1/2}$ evaluations of function $\mathcal{P} \circ E$. In order to evaluate the feasibility of this attack, it is necessary to estimate the size of $|\mathcal{P}(E(\{0,1\}^*))|$. The authors of *WalnutDSA* considered $q^{N(N-3)} \cdot N!$ a conservative lower bound for values of $\mathcal{P}$. For 128-bit and 256-bit security levels, these values were 2^{216} and 2^{336}, respectively, so it is expected to find a collision after 2^{108} and 2^{168} evaluations of $\mathcal{P} \circ E$. Hence, a generic collision attack is not practical.

In [20], it was shown (by means of computer experiments) that $|\mathcal{P}(E(\{0,1\}^*))|$ is at most q^{13} (lying in an affine subspace over $\mathbb{F}_q$), so a collision is expected to be found after $q^{13/2}$ evaluations of $\mathcal{P} \circ E$. With this new estimate, $2^{32.5}$ and 2^{52} evaluations of $\mathcal{P} \circ E$ are necessary for 128-bit and 256-bit security levels, respectively. Therefore, collision attack is practical in this case.

In order to implement this attack, the authors used a generic collision finding algorithm: the distinguished point algorithm of Van Oorschot and Wiener [24].

This algorithm finds collisions in any function $f : S \to S$ that behaves like a random function [24]. The time complexity for finding a single collision is $\mathcal{O}(\sqrt{|S|})$. A distinguished point is an element of S satisfying some easily testable property (e.g., a fixed number of leading zero bits). The distinguished point algorithm selects a starting point $x_0 \in S$ at random and produces a chain of points $x_i = f(x_{i-1})$ for $i = 1, 2, \ldots$, until a distinguished point is reached. Then, the starting point x_0, the distinguished point x_k, and the length of the chain are stored. It is expected that after $\mathcal{O}(\sqrt{|S|})$, the current chain will collide with one of the stored chains. Following the chain from that point, the same distinguished point will be reached.

In [20], this algorithm was applied to the function $f = g \circ \mathcal{P} \circ E$ instead of to $f = \mathcal{P} \circ E$, where g is a function that crafts plausible messages, given an output of $\mathcal{P}$. However, no implementation or description of how to build the function g was provided.

Using a standard PC, the algorithm found a collision after $2^{32.2}$ evaluations of f ($2^{32.5}$ evaluations were expected). This took approximately one hour. The two messages found by the algorithm were

$$\begin{cases} m_1 = & \text{"I would like to receive 7181666883746416503free samples of delicious cookies"}. \\ m_2 = & \text{"I pledge to donate 3519533052089988469 USD to Ward Beullens"}. \end{cases}$$

In order to mitigate this practical attack, Beullens and Blackburn [20] recommended to increase the value of q up to $q = 2^{20}$ and $q = 2^{40}$ to accomplish 128-bit and 256-bit security levels, respectively. With these new parameters, the size of the public key is five-times larger and the verification algorithm is 25-times slower for 256 bit.

A better mitigation of this attack is to change the encoding algorithm to output pure braids not restricted to the subgroup generated by $g_{N,1}$, $g_{N,2}$, $g_{N,3}$, and $g_{N,4}$. This change would require $q^{((N-2)^2+1)/2}$ evaluations of $\mathcal{P} \circ E$, and only a minor increase of parameters is needed. It was pointed out in [20] that a 256-bit security level could be accomplished by setting $q = 2^8$ and $N = 8$, making the key size 50%, the signature size 25% larger, and the verification algorithm two-times slower.

The authors of *WalnutDSA* pointed out that any braid output by the encoding mechanism E (see Section 4.1.2) is a product of the image (under $\mathcal{P}$) of the encoding braids used, and thus, it is essential that the subspace spanned by said images is sufficiently large [23]. They further depicted two design strategies towards defeating this attack (see Table 1).

Table 1. Examples of sequences to defeat collision attack (see [23]).

N	Periodic Sequence S	dim S	Recommended q
10	$\{(3,5,7,9),(2,4,6,8),(1,3,5,7),(2,4,6,8),\dots\}$	82	—
12	$\{(5,7,9,11),(4,6,8,10),(3,5,7,9),(2,4,6,8),(1,3,5,7),$ $(2,4,6,8),(3,5,7,9),(4,6,8,10),\dots\}$	122	$32,256$

5.3. Reversing E-Multiplication

A fundamental hard problem underlying the security of the *Walnut* signature scheme is to break the one-wayness of the function:

$$\mathcal{P}: B_N \to \mathrm{GL}_N(\mathbb{F}_q) \times S_N$$
$$s \mapsto (I_N, \iota_N) \star s.$$

Here, we write $\mathcal{P}$ instead of $\mathcal{P}_\mathcal{T}$ with the understanding that the set $\mathcal{T} \subset \mathbb{F}_q$ of non-zero elements is arbitrary, but fixed.

More precisely, the underlying problem is defined as follows.

Definition 8 (Reversing E-Multiplication (REM) problem [20]). *Given a pair $(M,\sigma) \in \mathrm{GL}_N(\mathbb{F}_q) \times S_N$, such that $(M,\sigma) = \mathcal{P}(s)$ for some braid $s \in B_N$, find a braid $s' \in B_N$ such that $\mathcal{P}(s') = (M,\sigma)$.*

Observe that if brute force is used to solve the REM problem, then it would take $\mathcal{O}(|\mathcal{P}(B_N)|)$ E-multiplications to find a solution, where $|\mathcal{P}(B_N)|$ is the size of the orbit of (I_N, ι_N).

Recall that the private key consists of two braids $s_1, s_2 \in B_N$, and the corresponding public key consists of $\mathcal{P}(s_1)$ and $\mathtt{mat}(\mathcal{P}(s_2))$, the matrix component of $\mathcal{P}(s_2)$. In [20], it was observed that a valid signature $\mathtt{sig}$ for a message m also satisfies:

$$\mathcal{P}(s_1) \star \mathtt{sig} = \mathcal{P}(E(H(m))) \star s_2. \tag{3}$$

Therefore, not knowing the permutation component of $\mathcal{P}(s_2)$ poses no problem to the attacker since it can be recovered from the permutation component of (3) without necessarily knowing the encrypted message (no message attack). Indeed, since cloaking elements and $E(H(m))$ are required to be pure braids, we have:

$$\sigma_{s_1} \sigma_{\mathtt{sig}} = \sigma_{s_2}.$$

Once σ_{s_2} has been computed, an attacker can solve two instances of the REM problem by finding two braids $s_1', s_2' \in B_N$ such that $\mathcal{P}(s_1) = \mathcal{P}(s_1')$ and $\mathcal{P}(s_2) = \mathcal{P}(s_2')$, which can be used to sign any message (universal forgery). Hence, solving the REM problem means that UF-NMA security (Definition 2) can be violated.

In this section, we describe two algorithms proposed in [20] that solve the REM problem. The first algorithm is a generic birthday attack, while the second exploits the structure of the braid group B_N and is more efficient than the first one.

5.3.1. Generic Birthday Attack

Given a pair $(M,\sigma) \in \mathrm{GL}_N(\mathbb{F}_q) \times S_N$, if we can find two braids $s_1, s_2 \in B_N$ such that:

$$(M,\sigma) \star s_1 = (I_N, \iota_N) \star s_2,$$

then the solution of the REM problem is $s' = s_2 s_1^{-1}$. In [20], it was argued that a naive way of finding s_1 and s_2 by constructing tables with values $(M,\sigma) \star s_1$ and checking if $(I_N, \iota_N) \star s_2$ for random s_2 lying in the table would take $\mathcal{O}(\sqrt{|\mathcal{P}(B_N)|})$ E-multiplications, making this method more efficient than a brute force approach. Nevertheless, a naive approach may require too much storage memory.

This inconvenience can be circumvented by using a distinguished point algorithm (see Section 5.2). In this case, the algorithm is applied to the function:

$$f(x) = \begin{cases} (I_N, \iota_N) \star s(x) & \text{if } b(x) = 0, \\ (M, \sigma) \star s(x) & \text{if } b(x) = 1, \end{cases}$$

where b and s are hash functions that take elements in the orbit of (I_N, ι_N) as input and output a bit or a braid, respectively.

The idea is to find collisions:

$$f(x_1) = f(x_2) \quad \text{such that} \quad b(x_1) \neq b(x_2).$$

Hence, if a collision is found such that $b(x_1) = 1$, then $b(x_2) = 0$ and $(M, \sigma) \star s(x_1) = f(x_1) = f(x_2) = (I_N, \iota_N) \star s(x_2)$. In this case, a solution of the REM problem is $s(x_2) s(x_1)^{-1}$. On the other hand, if $b(x_1) = 0$, then a solution of the REM problem is $s(x_1) s(x_2)^{-1}$.

As noted in [23], this attack is exponential in running time and can be thwarted by choosing the correct parameters for *WalnutDSA*, in this case $N = 10$, $q = 2^{31} - 1$ for 128-bit security, and $N = 10$, $q = 2^{61} - 1$ for 256-bit security.

5.3.2. Subgroup Chain Attack

This attack exploits the fact that the restriction of $\mathcal{P}$ to pure braids is a group homomorphism, which maps the chain of subgroups:

$$\{\iota_N\} = P_1 \subset P_2 \subset \cdots \subset P_N \subset B_N,$$

to a nice chain of subgroups of $\mathrm{GL}_N(\mathbb{F}_q)$. Here, P_k denotes the intersection of the subgroup of pure braids in B_N and the subgroup generated by $b_1, b_2, \ldots, b_{k-1}$, that is the subgroup of pure braids such that only the first k strands cross over each other. More precisely, for each $1 \leq k \leq N$, $\mathcal{P}$ is a homomorphism from P_k into the subgroup:

$$A_k = \left\{ \begin{pmatrix} X & Y & 0 \\ 0 & 1 & 0 \\ 0 & 0 & I_{N-k} \end{pmatrix} : X \in \mathrm{GL}_{k-1}(\mathbb{F}_q), Y \in \mathbb{F}_q^{k-1} \right\}.$$

In contrast to the birthday attack, this method solves the REM problem for a pair $(M, \sigma) \in \mathrm{GL}_N(\mathbb{F}_q) \times S_N$, by finding in iterative steps a braid $s \in B_N$ such that $(M, \sigma) \star s = (I_N, \iota_N)$, as follows. First, choose any braid $s' \in B_N$ such that $\sigma_{s'} = \sigma^{-1}$. Therefore, $(M, \sigma) \star s' = (M', \iota_N) \in A_N \times S_N$. Next, find a pure braid $s_N \in P_N$ such that $(M_N, \iota_N) = (M', \iota_N) \star s_N = (M, \sigma) \star s' s_N \in A_{N-1} \times S_N$. The iterative step consists of randomly choosing a target matrix $M_i \in \mathcal{P}(P_i) \cap A_{i-1}$ and then finding a pure braid $s_i \in P_i$ such that:

$$(M, \sigma) \star s' s_N s_{N-1} \cdots s_i \in A_{i-1} \times S_N.$$

Notice that in each iterative step, the permutation component is ι_N since s_i is a pure braid, and thus, $\sigma_{s_i} = \iota_N$. This process yields a braid $s = s' s_N s_{N-1} \cdots s_2$ such that $(M, \sigma) \star s = (I_N, \iota_N)$. Then, the solution to the REM problem is s^{-1}.

In [20], it was pointed out that if $M_i \notin \mathcal{P}(P_{i-1})$ for some $2 \leq i \leq N$, then it is not possible to complete the attack, and thus, assuming:

$$\mathcal{P}(P_i) \cap A_{i-1} = \mathcal{P}(P_{i-1}) \tag{4}$$

for each $2 \leq i \leq N$, guarantees that the attack will work. This assumption is not too restrictive since it seems to hold for the proposed parameters for *WalnutDSA*. With (4) in mind, the ith iterative step of this attack can be solved by performing a collision search in the space cosets of A_{i-1} in $A_{i-1} \mathcal{P}(P_i)$ with a cost of $\sqrt{|\mathcal{P}(P_i)|/|\mathcal{P}(P_{i-1})|}$ E-multiplications (see Sections 5.2 and 5.3 of [20] for details).

In [20], the running time of this attack was estimated to be $q^{N/2-1}$ whenever E-multiplication uses the set of invertible elements $\mathcal{T} = \{y_1, y_2, \ldots, y_N\} \subset \mathbb{F}_q$ with $y_a = y_b = 1$ for some $1 \leq a, b \leq N$ (see Section 4.2.2). It was noted in [23] that if y_a and y_b are chosen such that $y_a \cdot y_b = -1$, then the running time of the attack is increased to at least $\sqrt{x}\, q^{(N-1)/2}$, where $x = 60$ for $N = 8$ and $x = 96$ for $N = 10$. Moreover, this attack is defeated by taking $N = 10$, $q = 2^{31} - 1$ for 128-bit security, and $N = 10$, $q = 2^{61} - 1$ for 256-bit security.

5.4. Uncloaking Signatures

Kotov, Menshov, and Ushakov presented in [25] a powerful attack against *WalnutDSA*. It is a heuristic attack that works exclusively with braids and does not need to take into account E-multiplication. The authors reported experiments with one hundred random protocol instances with a 100% success rate. It is worth pointing out that the experiments were carried out for three different settings: the 128 and 256-bit security levels from the official specification [26] (where $N = 8$) and the 256-bit security version with $N = 11$, proposed in [27].

In a nutshell, the attack works as follows: An adversary, which collects several arbitrary pairs of messages and valid signatures, is able to compute an alternative secret key such that, when used to sign any message, it produces the same signature as the real secret key. Therefore, this is a very strong attack as it violates a rather weak security notion for signatures (UF-RMA; see Definition 3), that is an adversary with access to signatures for random messages (not adversarially chosen) can produce a valid signature for any message of its choice; that is, it achieves a universal forgery.

Next, we provide a high-level description of the attack:

- **Step 1.** The attacker collects k pairs $\{(m_i, \texttt{sig}_i)\}_{i=1}^k$ where each $\texttt{sig}_i$ is a valid signature for m_i computed with the same secret key (s_1, s_2). Each signature is a braid with the form:

$$\texttt{sig}_i = v_1^{(i)} \cdot s_1^{-1} \cdot v^{(i)} \cdot E(H(m_i)) \cdot s_2 \cdot v_2^{(i)}$$

  where $v^{(i)}, v_1^{(i)}, v_2^{(i)}$ are cloaking elements.
- **Step 2.** The attacker, using a heuristic procedure described in [25], is able to remove the cloaking elements from the signatures, that is compute braids $P_i = s_1^{-1} \cdot E(H(m_i)) \cdot s_2$. It is worth pointing out that Kotov, Menshov, and Ushakov reported a high success rate for their uncloaking algorithm, close to 80% or 100%, depending on the type of cloaking elements used (see Table 2).
- **Step 3.** The attacker computes the $k - 1$ products $P_i P_{i+1}^{-1}$. Note that these are:

$$P_1 P_2^{-1} = s_1^{-1} E(H(m_1)) E(H(m_2))^{-1} s_1$$
$$\vdots$$
$$P_{k-1} P_k^{-1} = s_1^{-1} E(H(m_{k-1})) E(H(m_k))^{-1} s_1$$

  obtaining a system of conjugacy equations in B_N where only s_1 is unknown. In [25], another heuristic algorithm to obtain a solution s_1' of the system (not necessarily equal to s_1) was developed.
- **Step 4.** The attacker sets $s_2' = E(H(m_i))^{-1} s_1' P_i$ for i of its choice. Under certain conditions, (s_1', s_2') works as an alternative secret key to (s_1, s_2), in the sense that it produces a valid signature for any message. Moreover, *as a braid word*, this signature equals the one produced with the original key. This implies that the attack cannot be avoided by limiting the size of accepted signatures. In order to decide if the alternative key (s_1', s_2') works as intended, Kotov, Menshov, and Ushakov generated signatures for 10 random messages and checked their validity.

Table 2. Percentage of properly-identified cloaking elements v_1, v, v_2 according to [25].

Encoding	Cloaking Elements	128-Bit	256-Bit	256-Bit with $N = 11$
Original	$wb_i^{\pm 2}w^{-1}$	80%	77%	76%
Original	$wb_i^{\pm 4}w^{-1}$	100%	100%	100%
Alternative proposed in [27]	$wb_i^{\pm 2}w^{-1}$	77%	81%	81%
Alternative proposed in [27]	$wb_i^{\pm 4}w^{-1}$	97%	98%	100%

In [25], a 100% success rate of the full attack was reported. One interesting fact is that the attack did not need many message/signature pairs in order to succeed: Kotov, Menshov, and Ushakov affirmed that, in all their experiments, six successfully uncloaked signatures were enough to get five conjugacy equations and a valid alternative secret key. Average running times for the full attack are shown in Table 3.

Table 3. Average running time (in seconds) for the full attack according to [25].

Encoding	Cloaking Elements	128-Bit	256-Bit	256-Bit with $N = 11$
Original	$wb_i^{\pm 2}w^{-1}$	18.8	120.8	213.0
Original	$wb_i^{\pm 4}w^{-1}$	17.4	112.4	185.6
Alternative proposed in [27]	$wb_i^{\pm 2}w^{-1}$	78.7	264.9	1674.9
Alternative proposed in [27]	$wb_i^{\pm 4}w^{-1}$	66.2	224.6	1323.3

With respect to possible countermeasures against their attack, Kotov, Menshov, and Ushakov themselves made several proposals. The first one is to artificially introduce many so-called *critical letters* in the secret braids (locating critical letters is one of the main ingredients in the uncloaking algorithm). In addition, they proposed using many more cloaking elements (around 30) on each side of the signature. Nevertheless, they pointed out that it is not even clear if this measure would be useful as it does not neutralize their attack [28] against Kayawood [29], another braid-based protocol. Finally, Kotov et al. recommended short conjugators for constructing cloaking elements, making them less visible.

The proponents of *WalnutDSA* recognize the weakness of their original implementation against the uncloaking attack and put forward in [23] a countermeasure against it. Namely, they introduced the concept of concealed cloaking elements and proposed to add six of them to the computation of each signature, which translated into a 6.7% increase of the signature size. Kotov, Menshov, and Ushakov questioned the effectiveness of the approach in the NIST PQC project discussion forum [27], pointing out that their algorithms were designed taking into account the existence of precisely three cloaking elements, but could be modified to deal with more of them.

6. Final Remarks

WalnutDSA is a beautifully-designed signature scheme, conceived in the remarkable mathematical scenario of braid groups. Despite the inspiring ideas involved in the construction of this scheme, the many attacks explained in this survey demonstrate that there is still a long way to go before a suitable key generation/parameter selection process is identified. We believe that it will be rather difficult to fix the security problems described, which may be an unavoidable consequence of the adept and symmetric signature procedure. A formal security analysis, as well as a deeper understanding of the actual relation between the cryptanalytic goals and the affiliated mathematical problems are essential ingredients for a secure implementation of *WalnutDSA*. Maybe a promising idea is to start by identifying the concrete cost of a forgery. For instance, a first step would be to assess whether a forger can be used in a black-box manner to reverse the related E-multiplication procedure (i.e., to solve the REM problem). Once such a result is at hand, the next step would be to look for solid instances of REM that could be used for secure key generation.

Author Contributions: All authors contributed equally to this survey, searching for related results, selecting relevant information and writing and reviewing the draft.

Funding: This research was funded by NATO Science for Peace and Security Programme, grant number G5448 and by MINECO under Grant MTM2016-77213-R.

Conflicts of Interest: The authors declare no conflict of interest.

References

1. Announcing Request for Nominations for Public-Key Post-Quantum Cryptographic Algorithms. Available online: https://csrc.nist.gov/News/2016/Public-Key-Post-Quantum-Cryptographic-Algorithms (accessed on 19 December 2016).
2. Persichetti, E. Efficient One-Time Signatures from Quasi-Cyclic Codes: A Full Treatment. *Cryptography* **2018**, *2*, 30. [CrossRef]
3. Hoffstein, J.; Howgrave-Graham, N.; Pipher, J.; Whyte, W. Practical Lattice-Based Cryptography: NTRUEncrypt and NTRUSign. In *The LLL Algorithm—Survey and Applications*; Nguyen, P.Q., Vallée, B., Eds.; Information Security and Cryptography; Springer: Berlin, Germany, 2010; pp. 349–390. [CrossRef]
4. Jalali, A.; Azarderakhsh, R.; Kermani, M.M.; Campagna, M.; Jao, D. Optimized Supersingular Isogeny Key Encapsulation on ARMv8 Processors. *IACR Cryptol. ePrint Arch.* **2019**, *2019*, 331.
5. Garber, D. *Braid Group Cryptography*; World Scientific: Singapore, 2007
6. Anshel, I.; Anshel, M.; Goldfeld, D. An algebraic method for public-key cryptography. *Math. Res. Lett.* **1999**, *6*, 287–292. [CrossRef]
7. Ko, K.; Lee, S.; Cheon, J.; Han, J.; Kang, J.; Park, C. New Public-Key Cryptosystem using Braid Groups. In *Advances in Cryptology, Proceedings of CRYPTO 2000*; Lecture Notes in Computer Science; Springer: Santa Barbara, CA, USA, 2000; Volume 1880, pp. 166–183.
8. Birman, J.; Gebhardt, V.; González-Meneses, J. Conjugacy in Garside groups I: Periodic braids. *J. Algebra* **2007**, *2*, 746–776. [CrossRef]
9. Katz, J. *Digital Signatures*; Springer: Berlin, Germany, 2010.
10. Goldwasser, S.; Bellare, M. *Lecture Notes on Cryptography*; MIT: Hong Kong, China, 2001.
11. Anshel, I.; Atkins, D.; Goldfeld, D.; Gunnells, P.E. WalnutDSA^{TM}: A Quantum Resistant Digital Signature Algorithm. *IACR Cryptol. ePrint Arch.* **2017**, *2017*, 58.
12. Artin, E. Theory of braids. *Ann. Math.* **1947**, *48*, 101–126. [CrossRef]
13. Anshel, I.; Anshel, M.; Goldfeld, D.; Lemieux, S. Key agreement, the Algebraic Eraser^{TM}, and Lightweight Cryptography. In *Algebraic Methods in Cryptography, Contemp. Math.*; American Mathematical Society: Providence, RI, USA, 2006; Volume 418, pp. 1–34.
14. Birman, J.S.; Cannon, J. *Braids, Links, and Mapping Class Groups, Annals of Mathematics Studies*; Princeton University Press: Princeton, NJ, USA, 1974.
15. Artin, M. *Algebra*; Prentice Hall: Upper Saddle River, NJ, USA, 1991.
16. Birman, J.S.; Ko, K.H.; Lee, S.J. A new approach to the word and conjugacy problems in the braid groups. *Adv. Math.* **1998**, *139*, 322–353. [CrossRef]
17. Dehornoy, P. A fast method for comparing braids. *Adv. Math.* **1997**, *125*, 200–235. [CrossRef]
18. Morton, H.R. The multivariable Alexander polynomial for a closed braid. In *Lower Dimensional Topology, (Funchal, 1998)*; American Mathematical Society: Providence, RI, USA, 2006; Volume 233, pp. 167–172.
19. Hart, D.; Kim, D.; Micheli, G.; Pascual-Perez, G.; Petit, C.; Quek, Y. A Practical Cryptanalysis of WalnutDSA TM. In *Proceedings of the Public-Key Cryptography—PKC 2018—21st IACR International Conference on Practice and Theory of Public-Key Cryptography, Rio de Janeiro, Brazil, 25–29 March 2018*; Part I—Lecture Notes in Computer Science; Abdalla, M., Dahab, R., Eds.; Springer: Berlin, Germany, 2018; Volume 10769; pp. 381–406. [CrossRef]
20. Beullens, W.; Blackburn, S.R. Practical Attacks Against the Walnut Digital Signature Scheme. In *Proceedings of the Advances in Cryptology—ASIACRYPT 2018—24th International Conference on the Theory and Application of Cryptology and Information Security, Brisbane, QLD, Australia, 2–6 December 2018*; Part I—Lecture Notes in Computer Science; Peyrin, T., Galbraith, S.D., Eds.; Springer: Berlin, Germany, 2018; Volume 11272, pp. 35–61. [CrossRef]

21. Merz, S.; Petit, C. Factoring Products of Braids via Garside Normal Form. In *Public Key Cryptography (2)*; Lecture Notes in Computer Science; Springer: Berlin, Germany, 2019; Volume 11443, pp. 646–678.
22. Paris, L. Braid groups and Artin groups. *arXiv* **2007**, arXiv:math.GR/0711.2372
23. Anshel, I.; Atkins, D.; Goldfeld, D.; Gunnells, P.E. Defeating the Hart et al, Beullens-Blackburn, Kotov-Menshov-Ushakov, and Merz-Petit Attacks on WalnutDSA (TM). *IACR Cryptol. ePrint Arch.* **2019**, *2019*, 472.
24. van Oorschot, P.C.; Wiener, M.J. Parallel Collision Search with Cryptanalytic Applications. *J. Cryptol.* **1999**, *12*, 1–28. [CrossRef]
25. Kotov, M.; Menshov, A.; Ushakov, A. An attack on the Walnut digital signature algorithm. *Des. Codes Cryptogr.* **2019**, 1–20. [CrossRef]
26. Anshel, I.; Atkins, D.; Goldfeld, D.; Gunnells, P.E. The Walnut Digital Signature Algorithm^{TM} Specifcation. Submitted to NIST PQC Project. 2017. Available online: https://csrc.nist.gov/Projects/Post-Quantum-Cryptography/Round-1-Submissions (accessed on 7 July 2019).
27. Comments to WalnutDSA^{TM} Proposal to NIST PQCProject. Available online: https://csrc.nist.gov/CSRC/media/Projects/Post-Quantum-Cryptography/documents/round-1/official-comments/WalnutDSA-official-comment.pdf (accessed on 7 July 2019).
28. Kotov, M.; Menshov, A.; Ushakov, A. Attack on Kayawood Protocol: Uncloaking Private Keys. *IACR Cryptol. ePrint Arch.* **2018**, *2018*, 604.
29. Anshel, I.; Atkins, D.; Goldfeld, D.; Gunnells, P.E. Kayawood, a Key Agreement Protocol. *IACR Cryptol. ePrint Arch.* **2017**, *2017*, 1162.

© 2019 by the authors. Licensee MDPI, Basel, Switzerland. This article is an open access article distributed under the terms and conditions of the Creative Commons Attribution (CC BY) license (http://creativecommons.org/licenses/by/4.0/).

MDPI
St. Alban-Anlage 66
4052 Basel
Switzerland
Tel. +41 61 683 77 34
Fax +41 61 302 89 18
www.mdpi.com

Symmetry Editorial Office
E-mail: symmetry@mdpi.com
www.mdpi.com/journal/symmetry

www.ingramcontent.com/pod-product-compliance
Lightning Source LLC
LaVergne TN
LVHW070626100526
838202LV00012B/737